Writing and Publishing Your Book

Writing and Publishing Your Book

A Guide for Experts in Every Field

Melody Herr

An Imprint of ABC-CLIO, LLC
Santa Barbara, California • Denver, Colorado

Library of Congress Cataloging-in-Publication Data

Names: Herr, Melody, author.
Title: Writing and publishing your book : a guide for experts in every field / Melody Herr.
Description: First edition. | Santa Barbara, California : Greenwood, [2017] | Includes bibliographical references and index.
Identifiers: LCCN 2017024019 (print) | LCCN 2017043402 (ebook) | ISBN 9781440858758 (hardcover : alk. paper) | ISBN 9781440859021 (softcover : alk. paper) | ISBN 9781440858765 (ebook)
Subjects: LCSH: Authorship. | Academic writing. | Authorship—Marketing. | Scholarly publishing. | Authors and publishers.
Classification: LCC PN146 (ebook) | LCC PN146 .H38 2017 (print) | DDC 808.02—dc23
LC record available at https://lccn.loc.gov/2017024019

ISBN: 978-1-4408-5875-8 (hardcover)
       978-1-4408-5902-1 (paperback)
       978-1-4408-5876-5 (ebook)

21 20 19 18 17   1 2 3 4 5

This book is also available as an eBook.

Greenwood
An Imprint of ABC-CLIO, LLC

ABC-CLIO, LLC
130 Cremona Drive, P.O. Box 1911
Santa Barbara, California 93116-1911
www.abc-clio.com

This book is printed on acid-free paper ∞

Manufactured in the United States of America

Contents

Introduction: Introducing the Expert

Experts like you are the reason I started reading serious books. As a teenager, I discovered the thrill of engaging on the page with experts who introduced me to Catherine the Great, explained chimpanzee behavior, and recounted the discovery of the double helix.

Experts like you are the reason I went to graduate school. As a college student, I relished conversations with Frank Bremer, the biographer of John Winthrop, and Steve Centola, the Arthur Miller scholar. When historian John Hope Franklin and paleoanthropologist Richard Leakey gave guest lectures, I arrived at the auditorium before the doors opened. Inspired by these face-to-face encounters, I applied to graduate school with the intention of becoming an expert and writing important books myself.

Experts like you are the reason I chose this career path. By the time I earned a PhD, I knew I couldn't write all the books that I believe really must be written. But as an acquisitions editor, I could assist other experts. For more than 16 years, I worked for a number of university presses, including Johns Hopkins and the University of Michigan. During that time, I published over 250 books in collaboration with authors in fields ranging from business history and U.S. history to law, political science, and international relations. Building on this experience, I moved into scholarly communications, an emerging profession that encompasses the economic, ethical, legal, and functional aspects of today's diverse, complex systems for disseminating research and scholarship. In this role, I help authors navigate these systems and make informed choices.

You are the reason I wrote this guide. You've invested years of study, research, and lived experience in becoming an expert. You know your topic, inside and out. However, writing and publishing a book require another layer of expertise, and that expertise is precisely what I offer. In the following pages, I show you how to construct a table of contents, prepare a proposal, identify a publisher, manage peer review, negotiate a contract, draft your manuscript, and market your book. In each chapter, I introduce the tried-and-true strategies that I developed through coaching hundreds of authors. These practical, step-by-step instructions will liberate your own creativity and keep you moving forward, efficiently and confidently.

Although I organized this guide from initial plan to final publication, it isn't necessary to read the entire guide before beginning your project. Maybe you're negotiating a contract or struggling through the first draft of your manuscript. Go ahead; jump straight to the relevant chapters. If you are just beginning, you'll benefit most from this guide if you read through it quickly to get an overview of the entire process and then work through the chapters one by one.

In Chapter 1, I discuss the role of books in the current information marketplace and then show you how to determine if a book is the best medium for your project using four interrelated criteria: methods, evidence, scope, and context. If you decide that your project meets these criteria, in Chapter 2, I help you to design your book so that the structure simultaneously supports and advances your argument. In Chapter 3, I assist you with assembling your book proposal—your expert portfolio—which includes your cover letter, CV, table of contents, and writing samples. As I explain in Chapter 4, the author–publisher relationship is a business partnership; you want to choose your partner with care. I enable you to evaluate potential publishers on the basis of their reputation in your discipline, the quality of their publications, and the level of service they provide both to authors and to customers.

After you've drawn up your short list, in Chapter 5, I explain peer review and the in-house evaluation process common to most university presses and commercial scholarly publishers. I propose strategies for approaching an editor, submitting your proposal, asserting yourself as necessary, and interpreting and responding to reviews. In Chapter 6, I walk you through the publishing contract, clause by clause, and explain the responsibilities of each party in this business partnership: the publisher and you, the author.

In the next two chapters, I teach you techniques for engaging readers even while you present them with complex, challenging material. In addition to these writing tips, I offer suggestions for using quotations, illustrations, graphs, and tables to maximize their impact as evidence for your argument. When you finish the manuscript, I help you prepare it for submission in Chapter 9 by explaining the components of a scholarly book and their arrangement. In the final chapter, I describe the publication process and give you a few pointers for collaborating with your publishing team.

Also in the final chapter, I show you how to launch a marketing campaign for your book. I help you to target the right audiences, with the right message, through the right media. Through your marketing campaign, you not only promote your book but also advance your career by spreading the word that you are *the expert* on your topic.

However you choose to use this guide, you'll find here the most essential information and the most effective strategies for writing and publishing your book.

ACKNOWLEDGMENTS

I want to thank the Greenwood team and, especially, Jessica Gribble for partnering with me to publish this book. Thank you also to the colleagues and authors who have worked with me and taught me so much. I look forward to future collaboration!

CHAPTER 1

Developing Your Project: The Criteria for a Book

You're making a name for yourself. You've won the recognition of colleagues at your institution and peers in your field. Now you're ready to solidify your authority and extend your domain. In combination with your other professional achievements, a book can be a highly effective means of establishing yourself as the go-to expert on your topic. In this chapter, I briefly compare a book to other venues for publicizing research and scholarship so that you can determine if a book will serve your purpose. I then discuss the four criteria for evaluating a potential book project: methods, evidence, scope, and context.

A BOOK IN THE INFORMATION MARKETPLACE

Pause for a moment and consider your own behavior as an information consumer. You seek information for answering a factual question, teaching a class, entertaining yourself during a dull moment, learning more about your research topic, and investigating larger questions in your field. News reports, journal articles, books, blogs, conferences, lectures, websites of professional societies, colleagues down the hall—all these are sources that you use at one time or another. How do your purpose and the nature of the information correlate with the source you choose?

Now take the opposite perspective. As an expert, you're an information producer. Most likely, you've already published selections from your project so you're familiar with the most common venues, their virtues, and their drawbacks. Blogs are popular because of their responsiveness, brevity, and extremely low cost; but they have a short shelf life and limited capacity for an evidence-based argument. Although a blog can be disseminated quite easily, the author's authority becomes diluted by commentary if, indeed, the author's name remains attached as the blog passes from device to device. Live presentations are constrained by time limits and the size of the auditorium, but they encourage give-and-take with the audience. Journal articles garner more respect and allow more space for argument and evidence, although even flagship journals seldom reach beyond their home discipline.

A book has more presence than any of these platforms. It makes a bigger splash and generates a wider, longer-lasting wake. The physical heft of a book is obvious. Whereas journal articles run to 10,000 or 12,000 words, nonfiction books average between 70,000 and 120,000 words, with books in the sciences and social sciences falling in the lower end of this range and books in the humanities in the upper end.

A book has more intellectual heft. Instead of a reaction to the headlines, it is an extended argument with supporting evidence, the product of reasoning and analysis. Its slow gestation from inspiration through research and writing to publication allows its author's thoughts to mature. It offers the insights of an expert. For this reason, a book remains relevant long after editorials vanish. Recognizing the enduring value of scholarly books, publishers are now converting their backlists, including titles published decades ago, to digital formats in order to keep them available.

For you, writing a book could propel your career to the next level by bringing you far-reaching, long-lasting recognition as the expert on your topic. A book gives your work more exposure than articles in specialty journals because a publisher has a vested interest in marketing it broadly, attracting as many customers from as many different fields as possible. A book has a price tag; but in addition to information, it offers the experience of a lengthy conversation with an expert. It is this experience that keeps a book fresh and appealing for several years.

Turning now to your current project, you can evaluate its suitability for publication as a book using four interrelated criteria: 1) methods, 2) evidence, 3) scope, and 4) context. I want to clarify that this is not a manual for revising a dissertation. The term *revision* misleads. Developing a book from a dissertation manuscript often requires not merely a careful housecleaning but a thorough remodeling. It may require altering the blueprints, knocking down walls, and uprooting plumbing in order to expand the foundations, raise a new framework, and build afresh. Nonetheless, if you've recently finished your dissertation and you hope to publish your research as a book, please keep reading. In the following pages, we'll evaluate the current shape of your project and discuss the design of those new blueprints.

METHODS AND EVIDENCE

Chemists aren't the only ones who study toxins in household cleaners. Meanwhile, performance studies scholars find research-worthy subjects in the pulpit, in spook houses, and in Civil War reenactment camps as well as on the stage. Traditional disciplinary boundaries dissolve as creative researchers and scholars embrace surprising new subjects, adopt new methods, and pursue new types of evidence. These trends overturn the conventional view that historians produce books, while economists write journal articles. Instead of discipline or topic, then, methods and evidence are factors for determining whether or not a particular research project warrants publication as a book.

The methods of narrative and critical analysis rely on interviews and ethnographic observations, personal letters and diaries, legislative records and court decisions, novels and newspapers, paintings and sculptures, music and poetry, reports on contemporary and historic events, films and television programs, maps, and photographs—the kinds of evidence that can be presented well in text and images on a page. If you use these methods and this type of evidence, your project meets the first criteria, so you can skip to the next subhead.

Quantitative studies are not as amenable to publication in book form if the numeric evidence appears in highly condensed, graphic summaries that only fellow experts comprehend. But if you do quantitative research, don't rule out the possibility of writing a book. Consider additional methods and evidence that you might incorporate. For example, a public health expert investigates the correlation between compliance with seatbelt laws and various demographic factors, including race, class, gender, education level, and family configuration. She harvests bushels of numeric data, runs analyses across several dimensions, and uncovers correlations from which she builds an argument. Government agencies, nonprofits, and auto manufacturers would welcome her research report, but book publishers probably would not because, as it stands, it consists of a scaffolding of data with only a few boards of text—sturdy and essential for a particular job but not comfortable or accommodating. And readers require comfortable accommodations in a book that they'll inhabit for several hours.

How can the public health expert make her study more inviting to readers and more attractive to publishers? She might conduct interviews with paramedics or police officers. When they arrive at an accident scene, which victims are wearing seatbelts? How do these on-the-ground observations compare with or supplement other data? She might examine public health warnings promoting the use of seatbelts. Are the signs in English or Spanish? Do they depict racial minorities? These questions lead her to textual and visual evidence. She could also look at the historic development of seatbelt laws and the public's response to them. She could investigate the evolution of car seats and seatbelt designs as well as the alarm systems reminding drivers to buckle up. These questions add a narrative dimension to her study. As a result, her project would satisfy the first two criteria for a book.

SCOPE AND CONTEXT

Scope refers to the dimensions of your project: the span of space and time as well as the depth of your investigation. Context refers to the broader world in which you situate a specific topic. Scope retains the focus on the same phenomenon, excavating it or tracing it to different locations. Context places the phenomenon under study within or alongside other phenomena that, at first glance, may not seem related. It's a matter of emphasis. Are you investigating one thing quite thoroughly, or are you examining how that thing relates to something else? Rather than speak in the abstract, I'll give you an example.

In her research on the leadership style of women legislators, a political scientist gathered quantitative data on elections, the demographics of legislatures, the configuration of committees, and the number and responsibilities of legislative staffers. In addition, she conducted interviews, analyzed legislators' speeches, and tracked the passage of a select number of bills. Although she started the project with Southern state legislatures, she expanded her scope to include the U.S. Congress and state legislatures in other regions. She also developed a comparison of women's leadership styles with that of their male peers. Going deep, she identified patterns in leadership styles correlated with race, party affiliation, number of years in office, and committee assignments. Now she could increase her project's appeal beyond the perimeter of her own specialty by situating her research in a larger context. But how will she choose the appropriate context—or contexts?

The most immediate context is the political ecosystem. Starting with the legislature, this researcher might describe floor rules, committee selection, and the legislative process. Moving outside the state house, she could survey electoral systems and party politics. Her examination of the institutional context would enable her to identify the leadership strategies likely to succeed in this ecosystem. In turn, her findings would invite a broader discussion about ways to promote diversity, civility, and cooperation within government. Political reformers will take note.

The institutional context, however, is not the only influence on legislators' leadership styles. The data reveal other contexts. What careers did the legislators pursue before they ran for office? Were they lawyers? Corporate executives? Independent business owners? Homemakers, nurses, or school teachers? What leadership styles do these various professions encourage? In the interviews, do the subjects mention inspiring role models, such as family members, fictional characters, or well-known politicians? How are women as leaders, in comparison to men, portrayed in contemporary movies and television programs? Illuminating the social context leads to exciting observations. In the conclusion, this political scientist might use her study of women legislators to discuss the ways in which the larger culture rewards or punishes women leaders.

This example illustrates the ways in which choices about emphasis and context affect a book's appeal to certain audiences. A focus on the institutional context limits the audience to fellow legislative studies scholars and, perhaps, a handful of reformers. A focus on the social context also welcomes gender studies scholars across the social sciences and humanities. Such a widely appealing book is also more likely to find its way into undergraduate classrooms.

If you're wondering who your target audience should be, start with the scholars in your neighborhood. They live next door, intellectually speaking; they study topics that overlap with your own. You study the U.S. Congress, they study the U.S. Supreme Court, and you encounter each other at the junction of judicial review. You study human–machine relationships, they study learning disabilities, and you meet in classrooms where innovative teachers use robots. Just as you don't invite the whole town to your backyard barbecue, you won't invite all your intellectual neighbors to your book. Sociologists interested in the ways in which architectural spaces affect human interactions also study the U.S. Congress, but you may have very little to say to them. Likewise, you may find electrical engineers incomprehensible, even when discussing a mutual obsession with robots. As you evaluate the many contexts in which you might situate your work, keep in mind whom you want to reach and select the context—or contexts—attractive to your target audiences.

THE NEXT STEP

If your project meets these four criteria—methods, evidence, scope, context—it has strong potential as a book. In the following chapters, I will show you how to write and publish a book that will draw attention beyond your own specialty and establish you as the go-to expert on your topic.

On the other hand, if you decide that a book doesn't suit your purpose or your project, you'll save yourself an enormous amount of frustration. Indeed, this decision opens other options. You can explore different venues for your current project or move on to a new project with better prospects for developing into a book. I'll share an example from my own experience.

As a graduate student, I wrote about a community of professional archae-ologists and hobbyists collecting fossils and human artifacts in the Midwest during the 1930s.[1] I gathered letters, photographs, field journals, scientific reports, and newspaper clippings. And I could chart social networks, map expeditions, and trace federal and state funding. I had, I thought, a fascinat-ing narrative, set in an austere landscape and populated with eccentric char-acters. But in addition to an entertaining story, I'd unearthed the culture of fieldwork. Associating fieldwork with a robust masculinity, professionals and hobbyists alike took pleasure in living off the land as they emulated 19th-century explorers or envisioned themselves as trophy hunters in pursuit of prehistoric animals. Indeed, the parallels with exploration and big-game hunting proved illuminating.

After I finished my dissertation—literally the day after I submitted it to my defense committee—I started working at Johns Hopkins University Press. Within a few weeks, I realized just how far my dissertation had to go in order to become a book. I had compelling evidence that I could analyze using sev-eral different methods, and I had situated my study in a context larger than that Midwestern community in the 1930s. But the scope was too limited. If another author submitted that manuscript to the press, I'd reject it.

With a full-time office job, I lacked the institutional support to complete the research for a book. I also lacked incentive. By that time, I'd decided to pursue a career in publishing and write books for young people. And so, I published one scholarly article; then, inspired by a collection of letters I'd found in the archives, I spun an historical novel for middle-school readers.[2] The decision not to invest more time and energy in the dissertation was the right one for me. Moreover, since that time I've had the privilege of assisting hundreds of talented experts with their books. Now I want to share what I've learned from that experience with you.

NOTES

1. Melody Herr, "Communities of American Archaeology" (PhD diss., Johns Hopkins University, 2000).
2. Melody Herr, "Frontier Stories: Reading and Writing Plains Archaeol-ogy," *American Studies* 44 (Fall 2003): 77–98; Melody Herr, *Summer of Discovery* (Lincoln: University of Nebraska Press, 2006).

CHAPTER 2 ———————————

Designing Your Book:
The Initial Structure

"What's wrong with alphabetical order?"

My friend sat across from me in the neighborhood café, watching the customers carry out their Sunday morning coffee rituals. I'd recently started working as an editorial assistant at Johns Hopkins University Press so I assumed she wanted professional advice. More likely, I realized years later, she simply wanted to boost my confidence. Whatever the case, after reading her proposal for a study of refugee women, I cared very much about her project; I was determined to help her write an unforgettable book. But that table of contents, arranged like a class roster, would launch it into obscurity.[1]

I asked why she'd chosen alphabetical order. Yes, it was logical. Yes, readers could easily find the woman whom they wanted to read about, although I pointed out that these women weren't famous and few people would recognize their names. Besides, weren't some of the interviewees identified by pseudonyms? My friend eventually admitted that the contents page reflected her filing cabinet. She'd organized her research by individual interviewees. Naturally she envisioned her book following the same order. And why, I pressed, had she adopted this filing system?

Her answer stopped my heart. Treating these refugees as raw material for an academic study demeaned them. They'd suffered violence and dislocation. Rather than dissect their stories, she wanted to keep them whole. I'd drawn diagrams to show her how to structure the book, but now I folded the sheets of graph paper and slid them into my tote bag. For the next three hours, we worked together to create a new table of contents that attended to her goals for the book, the reader's interests, and the refugee women's dignity.

Whether or not my friend wrote an unforgettable book, I will never forget that conversation. By forcing me to articulate what I'd learned as a reader and aspiring writer, it proved to me that I'd already developed the core skills required of an editor. (I continue to cultivate two other requisite skills: diplomacy and tact.) That conversation also demonstrated the dynamic synergy that results when an author and an editor collaborate. Since then, I've worked with and learned from hundreds of authors. Much to my regret, I couldn't take every one of them out for coffee, so I developed the following self-guided method. Now I want to share it with you.

YOUR ARGUMENT

You want to draw the blueprint for your book as soon as you've done enough research to determine that your project satisfies the criteria presented earlier in this guide. With this blueprint, you can identify the gaps in your research and avoid gathering material that doesn't fit. I often compare a book to a machine that transports readers. By giving them new knowledge, teaching them new skills, opening new experiences to them, urging them to act, or generally enlarging their perspective, a book moves them from one mental space to another. This machine consists of articulated modules, called chapters, which themselves are little machines built from several integral components. The engine powering this whole apparatus is the argument. Let's begin, then, with your book's engine.

Think back to the origin of your project. What questions enticed you to search for answers? What problems compelled you to seek solutions? As you conducted research and gathered information, you began to formulate an argument. An argument does more than answer the research questions or describe the findings. It persuades readers that you correctly identified correlations or cause-and-effect relationships.[2] In terms of the machine metaphor, *because* fuels your book's engine.

A description of the European Union's immigration laws sputters in the driveway. An argument that specific historical and social factors in established European democracies and former communist states determine the differences in their respective immigration laws roars down the highway. A chronicle of the Spanish Civil War (1936–1939) taxis, taxis, taxis. An argument about the ways in which certain events and persons prominent in the Spanish Civil War shaped the outcome of World War II zooms into the sky. From these examples, you can see how enlarging the scope of your research and situating it in a meaningful context converts a mere description into an intriguing argument. You can also see how scope and context define the target audiences. A scholar interested in the foreign policies of former communist states will pick up the book on immigration laws even if he's not interested in immigration per se. Military historians and military history fans fascinated by WWII will notice the book on the Spanish Civil War.

What's your argument? Write it down, incorporating its scope and context. Experiment with several versions—and keep them all because later you may want to return to them again.

DESIGN TEMPLATES

A book resembles a machine, propelled by an argument, assembled from modules called chapters. And the component parts of these chapters are the events, observations, data sets, facts, analyses, explanations, and conclusions that you produce in your research. Later I discuss how to assemble and name individual chapters. Here I focus on your book's overall design, which will eventually become your table of contents. I first present the method for designing a book in abstract terms. Then, in case you'd like a more concrete demonstration, I walk through a hypothetical example.

Nonfiction in any genre follows one of four templates for organizing content: 1) comparison/contrast, 2) theme, 3) chronology, or 4) geography. The comparison/contrast template identifies similarities between phenomena typically considered distinct or, vice versa, highlights differences between phenomena commonly thought to be similar. An author may discuss Phenomenon A point by point, then discuss Phenomenon B point by point in the same order:

Phenomenon A
  First Point
  Second Point
  Third Point
Phenomenon B
  First Point
  Second Point
  Third Point

Alternatively, an author may discuss the individual points, moving back and forth between A and B:

First Point
  Phenomenon A
  Phenomenon B
Second Point
  Phenomenon A
  Phenomenon B

The theme template allows an author to reveal the patterns that support his argument.

Theme #
  Argument and Evidence
Theme $
  Argument and Evidence
Theme &
  Argument and Evidence

The chronology template follows the order of time, while the geography template relies on the structure of physical space. An inexperienced author may misread *chronology* as *chronicle*. The two are not identical. Medieval monks wrote chronicles, running lists to the effect of this happened, that happened, this other thing happened. As a template, chronology simply supplies a guide for organizing content, allowing the author to recount and analyze events in the order in which they occurred. Instead of including every event and giving each one equal weight, the author selectively features the events that undergird his argument. In the same way, movement from place to place requires an itinerary that aligns with the argument's logic.

The templates provide an adaptable framework. Rather than strictly follow a single one, an author, journalist, or documentary filmmaker chooses a basic structure and modifies it to suit his material and purpose. Comparison/contrast may function best if the components are arranged geographically. A theme-based design may incorporate comparisons and contrasts.

When you sketch the blueprint for your own book, don't worry about designing a fancy table of contents. Right now you need only build a structure that supports and advances your argument. Start by writing each topic that you want to include (e.g., a character, an event, a place, an observation) on a separate index card. Then choose a template and experiment with different ways of arranging your cards according to this structure. Choose another template and experiment with other arrangements. Unleash your creativity! Repeat this process until you find a design with a clear internal logic that not only accommodates your engine but lets it run at full power. You would be wise to save all your notes from this session, including the dead ends, for future reference.

THE METHOD IN ACTION

In order to illustrate this method for designing a book, I use a hypothetical study of attitudes toward nuclear power in coal-producing states. Underscore *hypothetical*. Inevitably, an authentic expert on the subject will read these pages and discover bloopers; I apologize in advance. Given that I occasionally return to this example throughout the remaining chapters, let me take a moment to explain why I chose it. First, it meets the four criteria for a book. It offers various types of evidence, from interviews and government reports to maps and economic statistics; and it has a broad chronological and geographic scope. It also touches on many contexts. Equally important for the purpose of this guide, this subject is relatively noncontroversial. When I tested this section using concealed-carry laws and genetically modified organisms in conversations with colleagues, their intense response to these hot-button issues distracted them from the point of the exercise. Therefore, in an effort to keep the focus on the method rather than the example, I decided to use attitudes about nuclear power as an illustration. Please follow along as I demonstrate the method in action.

First, I write the best version of my argument on a large sheet of paper so I can keep it in view. "In the coal-producing states of Pennsylvania, Illinois, Kentucky, and West Virginia, from the 1940s through the Fukushima nuclear accident in 2012, perceptions of the risks of nuclear power depended upon the incidence of nuclear plant accidents, the influence of the coal industry, and the strength of the environmental movement." This argument takes into account the scope and context. My target audience includes sociologists and psychologists who study attitudes toward risk as well as historians who study the environmental movement.

Next, on index cards, I write the topics I want to cover: the environmental movement, coal industry statistics, nuclear industry statistics, key state legislators, leading lobbyists, Chernobyl, Three Mile Island, pollution. At this point, I don't worry about giving each card equal weight. Some cards bear huge, complicated topics, such as "health risks for miners"; others tote a small topic such as "Pennsylvania governor's speech." Nor do I worry about lumping topics together or splitting them apart (e.g., "Earth Day" is a subtopic of "Environmental Movement.").

Now the fun begins as I figure out how to arrange these topics into chapters and organize the chapters in a way that communicates the book's argument. I first try the geography template because I have four case studies—Pennsylvania, Illinois, Kentucky, and West Virginia—and this template allows me to tell the story of each state, complete with fascinating characters and dramatic scenes. Even as I sketch the outline, however, I realize I'm repeating background information in each chapter. Four times, I recite the history of the environmental movement as well as the history of nuclear plant accidents. Also, I risk losing readers in the details. In the introduction, they need substantial background information. In the conclusion, they need an explanation of how the stories prove my argument. Too much weight at the front and the back warp my machine's frame.

I rearrange the index cards and draft another design using the comparison/contrast template.

**Comparison/Contrast Template: Public Perceptions
of Different Types of Risks**

Introduction—overview of argument and issues
Personal Health and Safety, 1940s–1960s
   Attitudes toward Coal's Effects—accidents, miners' health—examples
      from each state
   Attitudes toward Nuclear Power—effects of radiation—examples from
      each state
Concerns about the Environment, 1940s–1960s
   Attitudes toward Coal—despoliation of the land, run-off pollution, air
      pollution—examples from each state
   Attitudes toward Nuclear Power—local waste; long-term, widespread
      radiation—examples from each state
Personal Health and Safety after Three Mile Island and Other Accidents,
   1970s–2012
   Attitudes toward Coal's Effects—examples from each state
   Attitudes toward Nuclear Power—examples from each state
Concerns about the Environment after Three Mile Island and Other Acci-
   dents, 1970s–2012
   Attitudes toward Coal—examples from each state
   Attitudes toward Nuclear Power—examples from each state
Conclusions—ways in which mining and nuclear accidents, as well as
   changes in cultural values, affect attitudes toward risk

This plan highlights the differences between fears about nuclear power
and those about coal. The overarching narrative supports my argument about
how growing awareness of health and environmental risks shaped these fears.
But the human-interest stories, gleaned from my case studies, must be pulled
apart to construct this narrative. I'm left with a well-structured but extremely
tedious design.

I try another design based on theme. You might point out that the previous
design follows two themes: environmental risks and personal risks. I agree.
The templates do not impose rigid structures but rather propose patterns on
which to draw an original design. If I center my design on the debate over
nuclear power, I can use vested interests, reformers, and at-risk populations as
my themes. I make cards for the vested interests: coal industry lobbyists, power
companies. I add a card for each of the reformers: leaders of the environmental movement, local civic groups. I include cards for the at-risk populations:
coal miners, residents of coal mining regions, nuclear plant workers, residents in the vicinity of a nuclear plant, anyone within reach of radioactivity
in the event of an accident. While making this list, I suddenly see that the

populations that consider themselves at risk change over the course of the 20th century. This is beginning to look like a book about interest group politics. But wait! This emphasis doesn't serve my argument.

I move on to the remaining template: chronology. An historian by nature and by training, I feel at home with chronological order. But immediately I'm overwhelmed. There are so many characters, so many events, so many facts in the complete history of the nuclear power debate in these four states from the detonation of the first atomic bomb through the Fukushima accident. Even worse, my argument resembles a hood ornament when it should be the force under the hood.

I pause to reread my argument. I flip through the topic cards. Then I spend some time studying the designs that I've already drafted. Gradually I rediscover the original motivation for this research project. I want to tell a story about the people who viewed themselves as victims or champions in a confrontation that would determine the fate of the economy, the planet, and humanity. The realization that I have more information about Pennsylvania than any other state leads me to notice the prominence of the Three Mile Island (TMI) nuclear accident. And there, at last, I find the frame for my story. Modifying the chronology template, I design a compelling narrative of the nuclear power debate in Pennsylvania.

Modified Chronology Template: The Pennsylvania Story

Introduction: TMI Accident (1979)—quick scene to grab attention; overview of the argument

Debate over Nuclear Power Prior to TMI (1960s–1970s)—focus on PA
   —coal industry in PA economy; mining accidents
   —growing concerns about personal health and the environment
   —key actors in the debate (vested interests, civic leaders, environmentalists)
   —refer to coal industry and debates in IL, KY, and WV for comparison/contrast
   Ideas about risk

TMI Accident—full story (1979)
   —focus on PA and local events; glance at national reaction

Immediate Aftermath of TMI Accident (early 1980s)
   —debate over nuclear power heats up; nuclear plant closures
   —focus on PA but use other case studies for comparison/contrast
   Ideas about risk

New Normal (late 1980–2010)
   —Chernobyl (1986) and reaction in the United States
   —fortunes of the coal industry

—fortunes of the environmental movement
—key actors; note changes in cast
—expand scope beyond PA; more about the other three states and the national scene
Ideas about risk
Fukushima Nuclear Accident (2012)
—reaction in the United States
—update on coal industry and nuclear power in PA; glance at situation in other states
Ideas about risk
Conclusion
—reemphasize the argument; highlight points from the narrative

As you see, I inserted the phrase *ideas about risk* as a reminder that the chapters must align with the argument. But listen to that argument again: "In the coal-producing states of Pennsylvania, Illinois, Kentucky, and West Virginia, from the 1940s through the Fukushima nuclear accident in 2012, perceptions of the risks of nuclear power depended upon the incidence of nuclear plant accidents, the influence of the coal industry, and the strength of the environmental movement." It no longer fits this elegantly designed machine, so I fine-tune it: "Over the past 60 years, nuclear plant accidents, the rise of the environmental movement, and the fortunes of the coal industry shaped perceptions of personal health and environmental risks, as illustrated by the story of the nuclear power debate in Pennsylvania." That's not perfect, I admit, but I will continually refine my argument as I write the book. Finally, I append a one-sentence abstract: "Focusing on the Three Mile Island nuclear accident, this study examines the debate over nuclear power in Pennsylvania from the 1960s through the 2010s and draws comparisons with the debate in three other coal-producing states: Illinois, Kentucky, and West Virginia." Later, when I work on the book proposal, I will add chapter titles.

MOVING FORWARD

Following this method, you can fashion a unique design for your own book. If you struggle, be patient. Save all your notes, all your dead-ends, all your potential designs. During the writing process, you may find they have unanticipated value. Give yourself a few days—a week if necessary—to experiment. When you need inspiration, study the contents pages of books in a variety of fields, particularly popular nonfiction. Then, whether or not you're satisfied with your best-so-far design, move on to the next chapter in this guide where I show you how to construct a book proposal and market your expertise to publishers.

NOTES

1. I've altered the details of this incident in order to protect my friend's identity.
2. For a quick review of how to formulate an argument, see Wayne C. Booth, et al., *The Craft of Research*, 4th ed. (Chicago: University of Chicago Press, 2016), 110–19.

CHAPTER 3 ────────────────

Writing Your Proposal:
The Expert's Portfolio

"Please send your proposal by e-mail. I'll look it over and get back to you within a week." Why had I answered the phone? In 20 minutes I had a lunch appointment downtown.

"No, you don't understand. I need to give my presentation in person. It's a big hit with my students. I tell political jokes, read satirical poems, and sing 19th-century American folk songs." The caller insisted on driving across the state in hope of persuading me to publish his book on Abraham Lincoln's foreign policy.

"January is a busy month for an editor. Please e-mail your proposal. I'm sorry: I have to run." When I returned to the office a few hours later, I expected the proposal in my inbox, but it didn't arrive until May. The author, so vivacious and enthusiastic, so imaginative in presenting his scholarship to live audiences, simply didn't know how to write a proposal. Judging from that proposal, I guessed he didn't know how to write a book either. What was he going to do? Give each reader a private performance?[1]

Like this eager author denied a stage, you have only the written page on which to perform. But that's all you need. Comparable to a professional artist's portfolio, your proposal displays your achievements and demonstrates your ability to undertake a successful new project—in this case, a book. In this chapter, you'll assemble a portfolio containing your cover letter, curriculum vitae (CV), table of contents, introduction, and a few writing samples. As elementary as it may seem, the structured approach that I present here serves two purposes. It makes your job easier because you know exactly what information to provide and how to arrange it. It frees you to concentrate on the creative, intellectual aspects of your proposal instead of wasting your energy guessing about the mechanics. At the same time, this structured approach makes the editor's job easier because you provide all the essentials in an orderly, digestible form. The more efficiently he can evaluate your proposal, the more likely he is to believe that you know what you're doing. No matter how ingenious your ideas, no matter how impressive your expertise, unless you demonstrate that you can produce a logical, sustained argument grounded in evidence and presented in an accessible way, an editor will not want to partner with you.

Keep in mind, however, that the acquiring editor alone doesn't decide whether or not to publish your book. Many individuals, from peer reviewers to marketing directors, participate in that decision. As I walk you through the portfolio, I'll point out who scrutinizes which sections and show you how to make your book proposal appealing to them.

Assembling a book proposal, like any process simultaneously demanding creativity and logic, is iterative. Be bold. Be patient. Above all, be playful.

YOUR COVER LETTER

"Dear Editor, here's my book proposal. I really hope you like it." All right, I never received a cover letter quite so crude, but how often I've seen one resembling a packing slip! As a university press editor bombarded by dozens of proposals every week, when I received such cover letters I felt as though I were a clerk in a corporate receiving department. I moved those proposals into a separate e-mail folder for later skimming. But once in a while I received a letter containing a concise, attractive description of the proposed book and the irrefutable credentials of the author. I find a good book—even the promise of a good book—irresistible. Enticed by the cover letter, I'd read the entire proposal immediately and, still wanting more, I'd give the author a call.

Your cover letter must seduce an editor. It counts more than any other part of the proposal, so don't treat it as an afterthought! Write it first, not last. There's an additional reason for starting with your letter: you'll use it as a guide for the rest of your proposal.

In this section, you'll draft a generic letter, which you'll later tailor for each publisher to whom you submit your proposal.

What do editors want? From my own experience as an editor and as an author, I can assure you that they want concise answers to five simple questions:

Who are you?
What's your argument?
Why does your argument matter?
What's your book about?
What's special about your book compared to other books on your topic?

These questions provide a guide for your cover letter. Introduce your argument. Assert its relevance. Describe the scope of your book. Highlight its unique aspects. Most important, throughout the letter, subtly display your credentials.

Returning to the example from the previous chapter, I drafted this cover letter for the book on the nuclear power debate.

SAMPLE GENERIC COVER LETTER: DEBATE OVER NUCLEAR POWER

DATE
Dear NAME OF EDITOR:

Today, the search for alternatives to fossil fuels centers on solar and wind power. Nuclear power rarely comes into the conversation; yet in the mid-20th century, its champions persuasively promoted it as an inexpensive, safe, clean energy source. When—and *why*—did Americans strike nuclear power from their list of options?

In the proposed book, I demonstrate that over the past 60 years, nuclear plant accidents, the spread of the environmental movement, and the fortunes of the coal industry influenced perceptions of the health and environmental risks of nuclear power. Focusing on the Three Mile Island nuclear accident (1979), I examine the debate in Pennsylvania from the 1960s through the 2010s and draw comparisons with the debate in three other coal-producing states: Illinois, Kentucky, and West Virginia.

Do we need another book on the nuclear energy controversy? Engineers identified the technical causes of the TMI accident, political scientists traced the development of nuclear regulations, and cultural historians documented the fear and fascination surrounding nuclear science. Yet none of these studies offer a close-up view of the controversy's chief actors as they responded to distant calamities as well as mishaps next door. Most important, none reveal how ordinary citizens evaluated the risks and benefits of nuclear power in terms of the jobs they needed to pay the bills, the cost of the energy they used to heat their homes, the safety of the water their children drank, and the beauty of the mountains where they hiked.

This is the man-on-the-street perspective that I offer. Drawing on interviews with miners and local residents, nuclear plant employees and coal company executives, state legislators and city council members, I capture the range of personal experiences. By searching energy industry statistics, national news stories, state legislative archives, and U.S. Nuclear Regulatory Commission records and reports, I place these personal experiences in the national and global context. Although I build on a sturdy historical framework and incorporate social science analyses, I construct a lively narrative with real characters engaged in conflicts with very high stakes.

Indeed, in Pennsylvania, one of the nation's largest coal producers and home to the nation's worst nuclear accident, the stakes were clear. For this reason, Pennsylvania provides an ideal focal point. But the conclusions and implications of my study reach far beyond the state's boundaries as the global debate over energy continues.

Along with this cover letter, I enclose my CV, a table of contents with a summary of each chapter, a draft introduction for the book, and one sample chapter. I welcome the opportunity to discuss this project with you. Thank you for your consideration.

Sincerely,
MY NAME
SIGNATURE BLOCK

If you read this example carefully, you'll see how I answered each of the questions. Baiting the hook in the first paragraph, I refer to the ongoing search for alternatives to fossil fuels and the mysterious disappearance of the nuclear power option. In the second paragraph, I introduce my argument and provide a one-sentence summary of the book. After acknowledging that researchers have written about nuclear power from many angles, I point to unanswered questions—the very questions that I answer in the proposed book. Although at first glance the fourth paragraph seems a mere list of sources, it also highlights my methods and my narrative approach. At the same time, it establishes my credentials as an author: I conducted interviews, collected statistics, analyzed news reports, searched through archives, and read the Nuclear Regulatory Commission records. Next, anticipating the criticism that I've written a limited case study, I assert the relevance of the Pennsylvania story both for the nuclear power debate in the United States and for the continuing global debate over energy sources. In closing, I list the contents of the proposal packet and thank the editor for considering it.

The cover letter functions as a sales pitch. It's your opportunity to advertise your book to the only customer who matters right now—the acquiring editor. For additional examples of seductive book descriptions, visit the websites of a few large American university presses. I recommend the University of Chicago Press, Johns Hopkins University Press, the University of California Press, and Harvard University Press, all of which employ full-time professional writers to produce the descriptions that make even the most esoteric books sound comprehensible and tempting. Collect eight or ten descriptions that catch your eye, print them out, and study them with care. Notice their common formula: the big issue or question, the specific topic, the features of the book, and the ramifications of the author's findings. Follow this formula as you write your cover letter.

YOUR CURRICULUM VITAE

Before moving on to your curriculum vitae, take one more look at the sample cover letter. In the fourth paragraph, the summary of the research subtly displays my authority. The information about me in my e-mail address, in my signature block, and on my letterhead also advertises my professional qualifications.

You, too, want to use every means to assert your expertise in a nonobtrusive way. You don't want to waste a paragraph describing your education, awards, career history, teaching record, and publications, although you might mention grants you received for the book project from high-profile sponsors. Pack these weighty details into your CV and write a sprightly cover letter. But pin down the corners with your professional insignia. I recommend that you use letterhead or, if you submit your proposal by e-mail, use a letterhead template that verifies your institutional affiliation. Also, if you submit your proposal by e-mail, use your institutional account. Close your letter with a complete signature block: your name followed by your degree, title, department, institution, postal address, institutional e-mail address, and office phone number. Please note that I'm using the terms *institution* and *institutional* to refer to your employer, which could be a university, corporation, government agency, nonprofit organization, or your own business—whatever professional affiliation verifies your credentials as an expert. If your employer insists that you use your official e-mail account and postal address solely for business purposes or if your current employment is not directly related to your area of expertise, you may use your personal contact information. I strongly recommend, however, that you choose an unambiguous, professional-sounding e-mail address. A moniker such as CallMeAl or FlowerKid will not create a favorable impression.

With the book description and the information contained in the conventions of a professional letter, you introduce your qualifications. In your CV, you enumerate them in detail. You already have two or three versions of your CV, each of which emphasizes your research, teaching skills, or another specific facet of your expertise. Now you want to create a version featuring your qualifications as an author: your academic credentials, your institutional affiliation, your research specialty, your publication record, and any other experience that uniquely qualifies you to write your book. Believe me—if you want to impress the acquiring editor evaluating your book proposal, efficiency matters more than eloquence. Be succinct.

At the top of your CV's first page, center your name and academic degree on the first line, your title and institution on the second line, your phone number and e-mail address on the third line.

Name, Degree
Title, Institution
Phone Number; E-mail Address

Returning to the standard margin settings, type "Highlights" in bold. Underneath, list the essential information about yourself in four or five bullet points. Consider your research, education, experience, awards, grants, significant publications, special skills. You've achieved so much, and you manifest your expertise in so many ways that you'll find deciding what to include difficult. But be selective; include only the achievements that speak to your expertise as the author of the proposed book. If it's about Chinese investment in North Africa, include "Fluency in Mandarin and French." If it's about marketing techniques, mention your own consulting business: "Partner, Three Marketeers." If it's about teaching youth the ethos of sustainability, absolutely list your Distinguished Service Award, Boy Scouts of America. On the other hand, if your book is about the Higgs boson, the Boy Scouts award does not belong here. This section should occupy no more than 20 percent of your first page.

After flashing your qualifications to catch the editor's eye, you can keep his attention fixed on your most significant qualifications, despite the details of your CV, by using three tricks: order of priority, bold headings, and reverse chronology. First, consider the categories for your achievements, for example, Employment, Education, Professional Service, Publications, Grants and Awards. Organize these categories according to their relevance to your book. Don't bury Publications at the bottom! I recommend that you place it in the third position, immediately after Employment and Education. As you did for Highlights, type each heading in bold, flush with the left margin. Certainly play with fonts and font sizes to satisfy your creative streak. Remember that design should enhance legibility and comprehensibility. Too many whirligigs pull readers' attention from the main attraction—your expertise! Finally, under each heading, list your achievements in reverse chronological order. By implementing this structure, you keep readers focused on your unique qualifications for writing this book.

YOUR TITLE

Since you first conceived this project, you've probably toyed with a number of titles; perhaps you found a favorite. But beware of witty titles! Potential readers may find them mystifying or downright distasteful. Quotations from the King James Version and Shakespeare, for example, do not play well. *Give onto Caesar* signals a book about taxes only to those familiar with Christian scripture. *The Taming of the Shrew* offends readers when they learn the book describes the proliferation of prescription antidepressants for housewives. Quotations from pop culture or your own sources, such as public speeches or personal letters, pose similar liabilities. There's an additional reason to forego a witty title: potential readers will discover your book primarily through database and online searches. Unless you put keywords—the right keywords—in your main title, a search may not capture your book or, if it does, the researcher may bypass your uncommunicative title in the list of hundreds of more revealing titles.

With these principles in mind, you can start devising your working title. Don't worry if it seems blunt or uninspired, as long as it serves the immediate purpose of advertising your book to publishers and peer reviewers. Later you can work with the publisher's marketing team to perfect it. Look over your argument and book blueprint, jotting down all the keywords. Then experiment with different forms (e.g., adjectives, gerunds, verbs) and different arrangements of these words. What combination works best? Which keywords belong in the main title announcing the large topic? Which belong in the subtitle clarifying and refining that large topic? A title for the study of women legislators, for example, might be: *Leadership Styles of American Women: Gender and Strategies in U.S. Legislatures.*

Returning to the example of the book on the nuclear power debate, I play with several potential titles. My keywords include nuclear, energy, reactors, fossil fuel, coal, mines; risk, danger, benefit, accidents, health, environment, radiation, pollution; jobs, economy, coal industry; change, long-term, tomorrow, future; citizens, lobbyists, government officials; values, contest, debate; Pennsylvania, Three Mile Island, coal-producing states. During the early days of the debate, contemporaries measured nuclear power's projected benefits and risks against coal's supposedly known benefits and risks. *The Devil You Know* tempts me—until I remember my own warning about cleverness. I save that phrase for a chapter title. For the book title, I try various keyword combinations.

Coal Mines and Nuclear Reactors: Health, Wealth, and the Environment in the Debate over Nuclear Power

This title is vague, although I like the health/wealth rhyme in the subtitle.

The Nuclear Power Debate: Changing Perceptions of Health and Environmental Risks in Pennsylvania

This version puts the subject squarely in the title and relegates the particulars to the subtitle. *Changing* indicates that I'm tracing developments over time. For emphasis, I add dates: *Changing Perceptions of Health and Environmental Risks in Pennsylvania, 1950–2015.* But I immediately remove them because they imply the termination not only of my study but also of my subject's relevance. Potential readers who discover the book in 2020 may dismiss it as useless history. Next I wonder if *Pennsylvania* is too specific, given that I present four case studies. What about *in Coal-Producing States*? That still seems too specific. For now, I simply shorten the subtitle: *Changing Perceptions of Health and Environmental Risks.*

Health and Environmental Risk Assessment in the Debate over Nuclear Power

Great, now I stuff all the keywords into a tongue-tiring main title and I can't think of a subtitle. This title conveys a static, uninhabited world, whereas I write a narrative with fascinating characters. And this thought leads me to two more variations.

Power Players in the Energy Debate: Coal Corporations, Nuclear Advocates, and Environmentalists
Debating the Risks of Nuclear Power: Coal Corporations, Environmentalists, and the American Public

Either of these titles might work, but neither quite aligns with my scope and argument. In the first title, the phrase *Energy Debate* implies that I cover the whole range of alternative energy sources when, in fact, I focus strictly on nuclear power. The term *Power Players* ignores the influence of nuclear accidents, which is essential to my argument. The title *Debating the Risks of Nuclear Energy* poses the same problem, although I like the inclusion of *the American Public,* which alludes to surveys and interviews. But maybe I should use the unambiguous term *Nuclear Energy* instead of *Nuclear Power,* which some potential readers might interpret as a reference to nuclear weapons as in "India has become a nuclear power."

Gathering what I've learned from each of these trials, at last I arrive at my working title:

The Nuclear Energy Debate: Changing Perceptions of Risk among the American Public

Keep experimenting until you land upon your own title. When you get frustrated, just choose the most accurate, most descriptive option. Keep in mind that your marketing team will help you to refine it later. And don't discard your experimental titles because they might ignite a future inspiration.

YOUR SIMPLE TABLE OF CONTENTS

In your proposal portfolio, I recommend that you include two versions of the table of contents: a simple version containing only the chapter titles and subtitles and an annotated version containing chapter summaries and information about your research. Start with the simple version and apply the same method used for generating the book title to the chapter titles. The only difference is that, now, tasteful wit is allowed. The table of contents should illustrate the progression of your argument. There's no need to repeat the topic in every chapter title because readers already know what your book is about. For example, the public health expert writing about compliance with seatbelt laws should not sow the phrase "seatbelt laws" throughout the table of contents. Rather than ask you to watch as I struggle with each chapter for the nuclear energy book, I'll simply show you the result.

The Nuclear Energy Debate: Changing Perceptions of Risk among the American Public

Introduction: Debatable Risks
Chapter 1 Coal versus Nuclear Power: The Devil You Know
Chapter 2 Three Mile Island: The Meltdown of Public Optimism
Chapter 3 After TMI: The Half-Life of an Accident
Chapter 4 Chernobyl: Those Reckless Russians
Chapter 5 Human and Environmental Health: Old King Coal on the Defensive
Chapter 6 Copiapó and Fukushima: Criminal Negligence or Acts of God?
Conclusion: The Relativity of Risk

Skimming my table of contents, the reader can see how I develop my argument. Whereas I restrict myself to keywords in the main titles, I indulge in wordplay in the subtitles. These playful phrases simultaneously make the topic sound intriguing and indicate each chapter's thesis. "Old King Coal on the Defensive" says the coal industry felt besieged. "Criminal Negligence or Acts of God?" acknowledges questions about the extent to which coal corporations and nuclear energy plants can prevent disasters. Let your own playfulness run free! On a well-organized contents page, the resulting balance of informative main titles and creative subtitles will both establish your expertise and exhibit your personality. People will want to meet such an authoritative and engaging author.

YOUR ANNOTATED TABLE OF CONTENTS

Please save your table of contents with a file name indicating that this is the simple version; then close the document. Reopen it and immediately save it again, this time with a file name designating it as the annotated version. By following the format presented here, you can concisely convey your book's structure and substance, your evidence and the ways in which you present it, and your progress with the research and writing. You also provide estimated word counts and the number of tables, graphs, and images so that the editor can calculate publication costs. Except for the summaries where you display your talent for distilling complex material, you don't need to write complete sentences. Remember, an editor values efficiency! Here's the recommended format:

Book Title and Subtitle
Introduction: Title
    Summary: [three or four sentences]
    Status: [status of the writing, e.g., not yet drafted, in progress, or drafted]
    Estimate: [word count (text and notes)]
Chapter 1: Title and Subtitle
    Summary: [one paragraph]
    Evidence: [short list of key sources for the chapter]
    Status: [status of the research, e.g., in progress, completed; also status of the writing]
    Estimate: [word count (text and notes); number of tables, graphs, images]
Repeat this structure for each chapter
Conclusion: Title and Subtitle
    Summary: [three or four sentences]
    Status: [status of the writing]
    Estimate: [word count (text and notes)]
Notes and References

Total Estimated Word Count (text, notes, references):
Total Estimated Visuals (tables, graphs, images):
Target Completion Date:

As an example, I use this format to describe one chapter of the hypothetical book on the nuclear energy debate.

Chapter 2 Three Mile Island: The Meltdown of Public Optimism

Summary: In March 1979, a mechanical malfunction triggered a meltdown at the Three Mile Island (TMI) nuclear power plant near Harrisburg, Pennsylvania. In this chapter, I recount the intense days as the world watched, fearfully wondering if the hydrogen trapped inside the disabled reactor would explode. When officials announced that the danger had passed and the radioactivity was contained, the national nuclear energy debate flared. For opponents, TMI foretold catastrophes to come. On the contrary, advocates pointed out, the worst predictions had *not* come true. TMI proved that the ingenious engineers who designed the facility had built a reliable system of safeguards. Pennsylvanians, shaken by the near miss, seemed to want only to shut down the plant and shut out the picketers.

Evidence: Contemporary newspaper accounts, interviews, and public opinion surveys; records of federal agencies (Department of Energy, Nuclear Regulatory Commission, Federal Emergency Management Agency); state of Pennsylvania executive and legislative records

Status: Research complete; draft chapter included in this proposal

Estimate: 15,200 words (text and notes); one graph; one drawing; six photos

As you prepare your annotated table of contents, you are not merely mapping out your book for an editor. You are also taking stock of your research, determining how to deploy your evidence, and organizing your content. With this well-designed strategy, your writing will progress more smoothly.

YOUR WRITING SAMPLES AND INTRODUCTION

Your professional writings, like a fashion designer's sketches, exhibit your expertise. What do you want to include in your portfolio? Although editors and peer reviewers generally prefer one or two draft chapters from your proposed book, you could provide journal articles, chapters from edited volumes, or conference papers. Whatever you select should relate to your book's topic, showcase your research skills, and illustrate your methodology. For instance, an expert on the rhetoric of the Chief Prosecutor of the International Criminal Court might include an article applying textual analysis to the Prosecutor's public speeches. If you choose unpublished writings, take care to proofread them. Delete every "note to self." Complete all the citations. Please don't submit a raw slide presentation or a lecture outline. A video will certainly draw attention, but use it to supplement—not replace—your writing samples.

When preparing your cover letter and CV, you appeal to the acquiring editor who judges the interest value of your project and your credentials. When choosing writing samples, you target peer reviewers, colleagues in your own field who evaluate the merit of your work. But when writing your introduction, you address the publisher's marketing team, a group of discerning readers who assess your ability to speak to professionals in neighboring disciplines and, depending on your topic, general readers. As much as they respect expertise, marketers want a product with sales potential beyond what they view as elitist disciplinary borders. Because you've already met the criteria enumerated earlier in this guide, you know that your book has wide-reaching potential. So how do you persuade the marketers?

The key to persuading readers of all sorts—especially marketers—is holding their attention. You excite their curiosity and then applaud interest in such a significant topic. While you exhibit your expertise, you never treat them as inferiors. Rather, with your charming writer's persona, you keep them plunging on, page after page, not only because of their unexpected fascination with your topic, but also because of their desire to learn about it from you. For your proposal, keep the introduction short, preferably under 3,500 words, and follow a formula similar to the one used for the description in your cover letter: 1) the attention-grabber, 2) the context and significance, 3) your argument, 4) your evidence and your study's distinctiveness, and 5) your findings and their ramifications.

The attention-grabber could be a dramatic scene, perhaps a recent event, an historic moment, or a glimpse of daily life, described in words evoking the senses. Indulge your inner novelist for two or three paragraphs. Invite readers to a legislator's campaign speech. Let them witness paramedics rescuing a pregnant woman from an overturned car. If you don't feel comfortable re-creating a scene, you can still make your attention-grabber visual. Recount the escalation of gun violence with year-by-year statistics or describe the AK-47, the preferred weapon of Mexican drug cartels. Moving quickly, place the scene in context and connect it to pressing issues. In one or two paragraphs, tell readers what it means and why they should take interest. Would seatbelts have saved the lives of the woman and her unborn infant? Do concealed-carry laws prevent or exacerbate gun violence? Given that the cartels obtain firearms from U.S.-based smugglers, should the Drug Enforcement Administration redirect its efforts to halt the drug trade?

The context creates an opening, ideally within the first 600 words, for you to insert your argument. State it concisely and then elaborate. Let it lead readers into the next few pages where you discuss the scope of your book and the extent of your research, emphasizing unique questions, sources, and methods. Expanding the description in your cover letter, highlight your excavation of antebellum state legislative records, innovative statistical analysis, interviews with nomadic chieftains in Central Asia. Editors and marketers care more about the "real-world" context of your work, but you may want to provide a literature review because peer reviewers expect a brief discussion of the current state of the field. In a few paragraphs, briefly describe the most significant comparable work, showing how you contribute to and build upon existing knowledge. Above all, without denigrating other writers, emphasize how your work surpasses existing literature, keeping in mind that these same writers might review your proposal; use diplomacy. You might, for instance, suggest that they simply did not have the opportunities that you do, thanks to newly available data or your special analytical skills. Next to your short evaluation of the essential literature, note the audiences within your own and neighboring fields for your book.

The editor, marketers, and peer reviewers will read your annotated table of contents—believe me, they will scrutinize it—so there's no need to reiterate it (e.g., "I cover this thing in Chapter 1; I talk about another thing in Chapter 2."). Nonetheless, by laying out your book's design in a subtle way, you reveal the progression of your argument. Instead of naming the chapters, use phrases such as: "I begin with . . . ; next I take a look at . . . ; moving on to . . . ; then I investigate; finally, I reach . . . ." Describing the progress of your argument eventually brings you to the book's conclusion. Conveniently, this is also the end of your introduction. In the last two paragraphs, summarize your key findings and point to their ramifications. Leave the reader poised to leap into the book.

MOVING FORWARD

By now you've noticed that this method for writing a book requires heavy intellectual and creative investment during the initial stages, but here you'll begin to see the profits. Just as your initial blueprint developed into your annotated table of contents and your cover letter generated your short introduction, all the material you assembled for your proposal portfolio will facilitate the next steps. Even your experiments with various book titles have potential value, so please don't discard them. You documented your expertise, you articulated your argument, you developed an innovative design for your book, and you persuasively explained its unique value. Now let's find a publisher!

NOTE

1. I changed the details of this incident in order to protect the author's identity.

CHAPTER 4 ———————————————

Investigating Your Publishing Options: The Search for a Partner

Writing a book can advance your career by bringing you far-reaching, long-lasting recognition as the expert on your topic. But what are your more immediate goals? Which audiences do you want to target? Do you need points for your tenure file? Do you covet a book prize? In order to meet these goals, you need the right publisher, a publisher with a team of professionals who understand your field and produce high-quality content in an attractive package. Equally important, you want a publishing team that treats you as a partner, simultaneously respecting your expertise and enabling you to do your best work.

In this chapter, I show you how to appraise prospective publishers. I begin with an overview of the different types of publishers and their respective business models. Then I help you to identify candidates and provide criteria for evaluating each one. The insights you'll gain by working through this chapter will empower you to make astute choices.

CLASSIFYING PUBLISHERS: BUSINESS MODEL AND MISSION

Nonfiction publishers come in four general types, defined by business model and mission: trade houses, commercial scholarly publishers, not-for-profit organizations, and university presses. Trade houses, such as Random House and Simon & Schuster, publish a mix of fiction and general-interest nonfiction (e.g., political commentary, lifestyle advice) by well-known authors. The trade model centers on a few high-volume bestsellers, supplemented by revenue from less popular titles, in each of the genres the house publishes. Trade editors rarely welcome a book for a specialized market, and they generally refuse unsolicited proposals. Consequently, if you want to approach a trade house, you need to hire a literary agent.

Although commonly associated with famous writers, some agents do represent experts like you. The Literary Agents Database, hosted by the nonprofit organization Poets & Writers and freely accessible online, lists a few nonfiction categories.[1] The annual *Guide to Literary Agents*, available in a print volume and by subscription to an online database, includes dozens of specific nonfiction topics in its drop-down search menu.[2] When you identify a potential agent, visit his website to find out which authors he represents and which books they've published recently. Does he have a successful record of placing books like yours with respected publishers? Agents offer a variety of services and fee schedules, so you want to make sure you understand exactly what you are paying for when you hire an agent. Likewise, even if you grant an agent responsibility for circulating your proposal and negotiating a deal with a publisher, you want to screen the publisher yourself, using the criteria presented in the next section of this chapter, before you sign the contract. Obviously, no agent can guarantee that a big-name publisher will accept your book. Nor can he guarantee that it will reach the bestsellers list, receive a golden review in *The New York Review of Books*, or earn you enough royalties to buy a waterfront condo in Chicago.

Like the trade houses, Greenwood, Routledge, and other commercial scholarly publishers publish across a range of disciplines. Textbooks and reference books, often with digital companions, account for a sizable portion of their revenue. At the largest publishers, the textbook and reference departments have dedicated editorial, production, and marketing staff. In order to balance these high-investment, high-risk ventures, these publishers produce a large number of high-priced, highly specialized books with smaller markets but fairly reliable sales. Small commercial publishers that focus on monographs, practical guidebooks, or supplementary books for the classroom in select disciplines follow a similar niche strategy. Lynne Reinner serves the social sciences, for example; Eerdmans concentrates on religious studies. As a result of this specialization, publishers understand their disciplines very well and excel at marketing to targeted audiences.

Whether you choose a corporation or a small publisher, generally you can send your proposal directly to an editor, and you are likely to find that his expertise aligns well with your own. A commercial publisher may not require peer review, but your proposal—and perhaps your finished manuscript—will be evaluated by a team of knowledgeable editors and experienced marketers who will appraise the quality of your work as well as its compatibility with current titles and its sales potential.

As you investigate commercial publishers, you may be surprised to discover that some charge the author for publication. A few of these so-called "vanity publishers" may conduct nominal peer review, but most require nothing more than a check or a credit card payment. This is good news if you want to avoid labor-intensive revisions and publish your book speedily. Self-publishing companies operate much the same way, although they usually offer a menu of services—copyediting, cover design, marketing—for additional fees. They also allow you to set the price and reap the revenue. In professional circles, however, vanity publishing and self-publishing are generally considered the last resort of a desperate author; they will not enhance your tenure file or your reputation as an expert.

Not-for-profit publishers exist to advance a mission, either service to a community or advocacy for an ideology or political agenda. Associations and societies publish books for their members. The American Library Association (ALA) serves professionals in library and information science, while the Minnesota Historical Society Press brings citizens books about their state. A few organizations, such as Brookings Institution Press, which supports scholarship on public policy issues, claim impartiality. Others audaciously promote their point of view. Hoover Institution Press, CATO Institute, and Templeton Press come to mind. In addition to publishing books, well-funded not-for-profits sponsor conferences, seminars, lectures, websites, magazines, and press releases—all of which feed into their overall marketing strategy.

Depending on your goals, a mission-driven publisher could serve your book well. Certainly, it would reach the community you want to address. These publishers excel at putting the right book in front of the right people. They know the audience, and the audience recognizes them as allies and sources of information. On the other hand, their range may not extend to neighboring fields. Does the ALA reach computer scientists, high school teachers, or social workers? You might also consider how your viewpoint aligns with that of an agenda-driven publisher. If you aim to make a bold ideological or political statement, you'll find ready sympathy. Yet be aware that although the like-minded will want to include your book in their arsenal because of the publisher's endorsement, opponents will deliberately ignore it for the same reason. Furthermore, academic peers and tenure committees at institutions committed to a belief in scholarly objectivity may frown upon such a publisher.

University presses endeavor to combine all of the benefits—and none of the drawbacks—of the other types of publishers. The ideal university press advances knowledge, culture, and social well-being as well as local pride *and* professes impartiality. It operates as a nonprofit, supported by its host institution, *and* attempts to generate enough revenue to maintain a degree of independence. It specializes in books in specific fields *and* strives for a catholic audience of academics, educated readers, and everyday people. You see the potential conflicts. Indeed, the effort to balance these challenges and aspirations affects every decision and every step of the publishing process, from peer review to pricing, from cover design to conference exhibits. An understanding of these inherent tensions will enable you, as an author, to make sense of the apparent contradictions and work successfully with the publishing team. That said, a university press may very well offer your best option: experience with books in your field, feedback from peer reviewers, and the endorsement of a respected institution.[3]

EVALUATING THE CANDIDATES

Now you're ready put together a list of potential publishers. You could start with a directory, such as the Subject Area Grid on the Association of American University Presses' website or the print guide and associated subscription database available from Writers Market.[4] Starting with your own bibliography or citation management system, though, would be more efficient. As you read through your citations, note which publishers crop up again and again. Note also which journals appear most frequently. For each journal, scan the book review section of several recent issues and jot down recurring publishers' names. Moving from the journal to its sponsoring professional society, open the webpage for book prizes. In the past five years, who published the winners? While you're on the society's website, click the link for this year's conference and read the exhibitors' roster. Repeat this three-step process—check the book review section, check the society's book prize winners, check the list of conference exhibitors—for the leading journals in your field and neighboring fields until you compile a list of eight or ten publishers.

With your newly acquired understanding of the different types of publishers, you can sort the candidates and give priority to the ones most likely to share your goals. Nonetheless, individual publishers within the same category may vary a great deal, so you want to investigate each with care. In order to guide your research, I provide a series of questions to evaluate a publisher's ability to serve you and to assist you in producing a book worthy of your name. You'll examine the publisher's commitment to your field and neighboring fields, the submission and review process, and the quality and appeal of the books it produces. Then, turning to marketing, you'll exchange the author's role for the reader's role. Like a secret shopper, you'll assess the publisher's website, check prices, scrutinize advertising, and inquire about distribution. Equally important, you'll interview recent authors about their experiences with the publishing team. Some publishers will drop off your list long before you reach the last question. Those that pass this rigorous investigation deserve your consideration.

BOOKS AND SERIES

Does the press publish books in your field? How many titles appeared in the past two years?

If the press hasn't published many—or any—new books in your field in the past two years, possibly the editor decided to discontinue this subject area, in which case he will not welcome your proposal. Alternatively, your field may be a new direction for the press and he is eagerly recruiting new titles. A decline or an increase in the number of books published annually over the past few years indicates which scenario applies.

Do any recent books cover your topic?

Whereas two or three books indicate interest in your topic, after a certain point the publisher will turn away a new one that might compete with its current titles.

Does the press publish series in your field? Do you recognize the series editors? What are their credentials?

Although scholarly publishers offer standalone titles, when an acquisitions editor invites highly regarded scholars to serve as series editors, everyone benefits. The publisher benefits from the experience of the series editors, who recruit authors, recommend peer reviewers, mentor junior colleagues, and generally advise the acquisitions editor about trends in the field. Scholars benefit from a wellspring of books. You, the author, benefit from the series brand and continual marketing because the publisher will include your book every time it advertises new titles or the series as a whole.

Speaking as a former acquisitions editor, I'll confide that a proposal or a manuscript submitted for a series often glides through peer review. As a collegial gesture to the series editors, leaders in the field volunteer as reviewers and, because of these professional relationships, they tend to turn in their reviews promptly. The series editors' endorsement carries weight with the publisher's in-house editorial committee and, at a university press, with the faculty advisory board. Then the book rides smoothly through production because copyeditors and designers follow the series routine. When the book reaches the warehouse, marketers add it to the ongoing promotional campaign for the series. All told, publishing in a series improves efficiency and reduces the aggravation potential throughout the process for both the publishing team and the author.

The very attributes that make a series ideal for some authors, however, make it a less desirable option for others. A marketing professional may never admit that branding has drawbacks, but it does. From packaging to intellectual content, a book series is a brand. No matter how open-ended its name, it does not merely cover a subject; it also presents a methodology or theoretical stance. The better your work aligns, the better the series enables you to speak to like-minded peers—while it simultaneously constricts your efforts to reach beyond this circumscribed group. The limitations of a series situated in a particular discipline should especially concern you if you do interdisciplinary work. Sociologists and criminal justice scholars, for example, may not discover a book published in a psychology series. In the same way, the endorsement of the series editors may help you secure a publishing contract, but their names on your book may prove a liability in other situations. Finally, the series routine, which expedites the publishing process, precludes choices about the cover and interior design for an individual book. Yours will resemble others in the series. Only you can determine whether or not a series will advance your goals.

Have recent books won major prizes in your field?

Prizes attest to the quality of the publications as well as the marketing team's knowledge of your field and ability to place books in the right hands.

Do the leading journals and websites in your field review books published by this press?

As with prizes, the fact that journals and websites review these books indicates the publisher's understanding of your field. But take time to read some of the reviews. What do reviewers say about scholarly merit? Do they comment on the copyediting or production quality?

Turn now to the neighboring fields that you want to address and repeat all of these questions.

SUBMISSION AND REVIEW

Which acquisitions editor manages books in your field?

When you locate the press directory, you will find half of the staff have the word "editor" somewhere in their titles. If I answered "I'm an editor," when a fellow passenger asked me what I did for a living, I spent the next half-hour explaining what I did not do. I was not a news editor, calling the shots at a city paper. I was not a copyeditor, proofreading and correcting grammar. I was not a developmental editor, sanding out a manuscript's rough patches, or a ghost writer, scribbling memoirs for Hollywood celebrities. I was not even a managing editor, guiding the books in my charge through the production process. I was an acquisitions editor, an intellectually curious entrepreneur who strategically recruited top-quality authors in order to publish a competitive line of books in select fields. (As you begin contacting acquisitions editors, you will discover how many share this self-image; use it to your advantage.) Not surprisingly, after this explanation, my fellow passenger slept quite well throughout the rest of the flight.

You are looking for the acquisitions editor who recruits authors in your field. Just as a professor in large university department can teach very specific courses (e.g., "Renaissance Italy" or even "Women and Religion in Renaissance Italy"), an acquisitions editor at a large press may focus on a specific discipline or subdiscipline, whereas his colleague at a smaller press may cover several areas (e.g., Humanities or Music & the Arts). In case you aren't certain which editor to approach, choose one and only one, trusting that if you've guessed incorrectly, he will route your proposal to the right person. Sending your proposal to multiple editors will result in confusion and miscommunication. Sending your proposal to the press director (unless he also acquires books in your field) will not win you points; on the contrary, it may delay a response.

Does the acquisitions editor have academic credentials and publishing experience?

Once you've identified the acquisitions editor, check his credentials. Does he have a graduate degree? Ideally, he will hold a degree in your field or a neighboring field. But regardless of the discipline, graduate training provides an understanding of research and scholarship, so he will have some idea of the rigors of your work and the value of your expertise. How much publishing experience does he have? Has he worked for other presses? How long has he worked for this one? In what fields has he acquired? You're looking for an editor whose expertise complements your own.

Does the press post submission guidelines on its website?

Although you'll prepare your proposal for submission in the next chapter, right now you want to make sure that the press does accept submissions directly from an author.

Does the press conduct peer review? If it is a university press, does it have an executive committee or a faculty advisory board that must approve the manuscript before publication?

The press should outline clearly the process by which it evaluates proposals: in-house review, peer review, and, if applicable, review by an executive committee. Which levels of review does it require in order to offer a publishing contract? Which additional levels does it require before it publishes a manuscript? Approximately how long does the review process take?

DESIGN AND PRODUCTION

Packaging matters. No matter how stimulating and significant the intellectual content, readers also want a well-crafted book. From cover design to copyediting, the product should reflect the expert quality of your work.

Do you find the book cover designs attractive?

Look at thumbnails on the website or, better, walk past the exhibit booth at a conference. Do the book covers catch your eye? Do the images bait your curiosity? Can you read the title and the author's name at a glance? Can you comfortably read the rest of the type?

Does the interior design complement the style and quality of the contents?

Most readers do not notice the interior design, unless it interferes with their comfort. Illegible type, odd page breaks, narrow outside margins, text dribbling into the gutter—these features suggest sloppiness on the part of the publisher and, as a consequence, reflect poorly on the author. On the other hand, a designer may use an unconventional font and a novel layout in a book about modern art or poetry in order to visually enhance the author's argument.

Does the press provide competent copyediting?

Here again, the readers' impression of the packaging is inextricably correlated with their impression of you, the author, the expert who wrote the intellectual content. Take time to look at several books published by this press. Can you confidently entrust your manuscript to this copyediting team?

MARKETING AND DISTRIBUTION

Switching roles, you'll now evaluate the press as a customer. You want to determine how the press assists readers in discovering books of interest.

Does the website, overall, feel inviting? Is it appealing or drab? Is it too busy or too sparse?

You may not be a web designer, but you instinctively know if the website makes you want to stay awhile.

Is the website well organized? How easily can you find books in your field and in neighboring fields? How easily can you find series?

Whereas a customer will likely head for an online retailer in search of a particular title, he will browse a publisher's website for a given discipline or topic. Does the website replicate the experience of walking through a bookstore?

How well does the search box function? How difficult is it to find a specific title or author?

If a customer does come to the site for a specific book—your book, for example—the search function should take him directly to it, even if he misspells the author's name or mangles the title.

How much information does an individual book's webpage provide?

At a minimum, a book's webpage should contain the same information as a book cover: the title, the author's name, a description, a few promotional endorsements (aka, blurbs), the name of the series (if relevant), and a note about the author's credentials. Preferably, the webpage also includes a table of contents and an excerpt from the text. Following the retailers' lead, publishers have begun listing a few "also of interest" titles in the margin of a webpage. This feature will help to direct readers to your book.

What formats does the press offer for books in your field? Hardcover? Paperback? E-book? Do individual books appear in multiple formats? What does the press charge for each of these?

A press develops standard format options and pricing structures, determined by the field and the audience. A paperback for undergraduates about Franklin D. Roosevelt's administration will cost less than a hardcover for legal scholars about criminal justice in South Africa. Familiarize yourself with how the press handles books for your intended audiences. Take a few minutes, too, to investigate the different types of e-books offered and the distribution channels for them. If you assume that because the press publishes e-books, yours will automatically appear in a Kindle, Nook, or iBook edition, you could be disappointed.

Does the press offer exam copies for instructors? How readily can you find this information?

As you know, an exam copy can influence an instructor's decision about assigning a book to her students. If you're writing a book for the classroom, you want to check if the press makes exam copies available and if potential instructors can quickly find the form for requesting them. A press catering to instructors will put a link to the form directly on a book's webpage.

How many marketing strategies does the press employ? How effective are they?

Find out which conferences the press attends and make a point to visit its booth. Are the staff knowledgeable and excited about their books? Is the booth well organized? Are the books in good condition? An exhibit, from the representatives manning the booth to the display furnishings, reveals much about the quality of the staff and the management of the press as a whole. Also, because of the costs involved, the decision to host a conference booth indicates the press's commitment to the field.

Check the website for subject catalogs. Does the press publish enough books in your field—and care enough about it—to issue a special catalog? Does this catalog list only recent titles, or does it include somewhat older titles? Consider how long after publication the press will continue marketing your own book.

Register for the press newsletter or mailing list. Does the press send out a single, one-size-fits-all e-mail or does it feature lists devoted to your field or neighboring fields? When you start to receive messages, are they compelling? Although publishers are stepping away from print advertising, you want to scan journals and conference programs. "Space ads," as marketing professionals call them, can be highly effective with a captive audience seeking something to occupy their attention. How many times have you read posters while riding the subway? How often have you paged through a conference booklet or a handful of flyers while listening to a presentation or waiting for a colleague?

Search the review sections of leading journals and websites. Do press titles appear regularly? The marketing team harvests quotations from reviews and posts them on the book's webpage, so a glance at the credit lines under these blurbs will indicate where the book was reviewed.

Look at social media hosted by the press. The number and type of social media accounts matter less than how frequently the marketing team refreshes them to keep consumers coming back. Do the posts report happenings at the press, discuss current affairs, or introduce new books? Do authors participate in social media? Does the press sponsor a blog? Who writes content for it?

Finally, check the "news" section of the website where the press announces prize winners, book signings, radio interviews, lectures, and other author events. Does the press take pride in its authors' achievements? How much does it support their marketing efforts? Can you imagine yourself promoting your own book in these ways?

Even before you submit your proposal, this comprehensive survey of the publisher's marketing efforts will give you a good sense of how actively—and astutely—it works to promote a book and the expert who wrote it.

Does the press operate distribution systems in Asia and Europe or does it contract with a distribution company?

Your audience extends beyond the United States; the publisher needs to ensure that the availability of your book extends equally far. Information about distribution is usually found under the "contact" section of a press's website. Scrolling down to the sales division, look for staff responsible for foreign accounts. Go ahead and contact these individuals (or dial the anonymous phone number or write to the anonymous e-mail address). At first, the person who responds may mistake you for a retailer; but once you identify yourself as a potential author, he should answer your questions courteously. If he doesn't, you will have learned something important about the way the press treats clients.

Does the press offer open access options?

An increasing number of authors seek to distribute their work in digital formats offered on the web without charge to readers. Known as "open access," this form of distribution has become a common option for articles published in scientific journals. Publishers are testing various funding models that will enable them to offer this option to book authors as well. For example, a number of university presses participate in Knowledge Unlatched, which makes a selection of high-profile titles available on the web. Meanwhile, other university presses are experimenting independently with open access. Amherst College Press publishes all of its titles in open access formats; the University of California Press created the program Luminos. If you received a grant from an organization that requires open access for publications resulting from your research and if you're able to supply a subvention to underwrite the cost of publication, you may be able to negotiate with your publisher of choice for open access distribution.[5]

THE AUTHOR'S EXPERIENCE

Online retailers invite customers to write reviews of products and services. To my knowledge, no publishers collect similar reviews from authors; you'll need to interview a few yourself. Pleased authors and, conversely, extremely unhappy authors will be the most eager to talk with you. Merely satisfied authors, like customers who get exactly what they expect, probably won't have much to say. Start with colleagues you know, preferably those who have published multiple books so they can compare the different presses with which they've worked. In case you don't know anyone who has published with a particular press, go ahead and reach out to authors in your field who have. Here

I pose a few questions to start the conversation, but certainly you'll have your own questions; and the interviewee will want to share the highlights of his experience.

How well does the publishing team understand your field?

Obviously, you want a publishing team that understands your field, acknowledges your expertise, and values your book's contribution.

Did the publishing team treat you with professional collegiality? When you disagreed with their suggestions or when problems arose, how did they react?

The publishing team should treat the author as a partner in the business of producing a well-written, well-designed, highly marketable book. At the very least, the team should keep the author informed of decisions and the reasons for them.

After you submitted your proposal, how long did you wait to receive reviews? Did the acquisitions editor discuss the reviews with you? When did the press offer you a contract?

You want to know how many months the review process may take so you can plan accordingly. You also want to inquire about the acquisitions editor's level of engagement. For example, did the editor communicate with your interviewee regularly while they were waiting for the reviews and work with him to write a response? How complicated and time consuming was the approval process? Your interviewee might not feel comfortable talking about his contract negotiations, but you could walk to the edge of that topic and see if he wants to pursue it.

After you submitted the final manuscript, how long did production take? What challenges did you face during the process?

Here again, if you have a publication target date, you need to know how much time to allow for production and whether or not the publishing team keeps an author apprised of his book's progress.

Would you say that the marketing team reached the audiences that you wanted to address? What marketing strategies did they use? To what extent did the press encourage and support your own marketing efforts?

Ask specifically how your interviewee envisioned his readers. Did he aim for his field or beyond? Which audiences did the marketing team target most effectively? Also consider your interviewee's tendency toward introversion or extraversion; some authors prefer to leave publicity to the press, while others adore the spotlight.

Is there any additional advice you would give an author?

People love to give advice! Invite your interviewee to pass along words of wisdom.

GOALS AND PRIORITIES

After this exhaustive investigation, revisit the goals that you listed at the outset. If you've identified a publisher who meets your each and every desideratum, congratulations. More likely, you didn't discover the ideal publisher and you need to determine which goals you most value and which you are willing to compromise. Maybe a series brand means more to you than a unique book design. You might feel compelled to choose publishers favored by your tenure committee. Alternatively, a colleague's experience with a small press may have convinced you that you want the same personal attention. Regardless of the factors affecting your decision, the information you collected empowers you to make astute choices now and to negotiate a book contract strategically later. With your short list of prospective publishers in hand, you're ready to move on to the next chapter and develop a plan for approaching them.

NOTES

1. Poets & Writers, Literary Agents Database, accessed 24 February 2017, https://www.pw.org/literary_agents.
2. Chuck Sambuchino, ed., *Guide to Literary Agents 2017: The Most Trusted Guide to Getting Published* (New York: F+W Media, 2016). Writers Market, Category Search—Literary Agents, accessed 24 February 2017, http://www.writersmarket.com/MarketListings/LiteraryAgents/search.
3. The Association of American University Presses does an excellent job of explaining the mission of a university press as well as the challenges it faces; accessed 27 December 2016, http://www.aaupnet.org/about-aaup/about-university-presses.
4. Association of American University Presses, Subject Area Grid, accessed 24 February 2017, http://www.aaupnet.org/images/stories/documents/aaupsubjectgrid2017.pdf. Robert Lee Brewer, ed., *Writer's Market 2017: The Most Trusted Guide to Getting Published* (New York: F+W Media, 2016). Writers Market, Category Search—Book Publishers, accessed 24

February 2017, http://www.writersmarket.com/MarketListings/BookPub
lishers/search.

5. Knowledge Unlatched, accessed 27 December 2016, http://www
.knowledgeunlatched.org/; Amherst College Press, accessed 27 December 2016, https://acpress.amherst.edu/about/; University of California
Press, accessed 27 December 2016, http://www.ucpress.edu/openaccess
.php. For an example of an open access book sponsored by a funding
organization, see Raymond J. La Raja and Brian F. Schaffner, *Campaign
Finance and Political Polarization: When Purists Prevail* (Ann Arbor: University of Michigan Press, 2015), accessed 27 December 2016, http://quod
.lib.umich.edu/u/ump/13855466.0001.001.

CHAPTER 5 _____

Launching Your Charm Offensive: Proposal Submission and Peer Review

That Monday morning, returning to the office after a week-long vacation, I dreaded opening my e-mail. Among scores of institutional communications, announcements from professional societies, and industry news alerts, I found 27 book proposals, most of them carrying the ambiguous subject line "Book." As an acquiring editor at a mid-size university press, I was managing approximately 30 titles under review and in production, plus an equal number of prospects in development, plus four or five times that many published books. Yet each of those 27 authors wanted me to focus on just one project— his own.

With that kind of competition, how do you get your proposal noticed? And once you do, how do you persuade the acquisitions editor that you're a conscientiousness and amiable colleague as well as a first-class expert? You need a charm offensive, a strategy calculated to attract attention and garner favor. In this chapter, I offer a tactical plan, based on my professional experience and innumerable conversations with publishing colleagues, for winning your way through the evaluation process. I begin by showing you how to submit a standout proposal. Next, I describe peer review and enable you, the author, to take an active role in the process. I help you to anticipate common review scenarios and prepare a persuasive, well-reasoned response. Lastly, I describe the deliberations of the publisher's in-house editorial committee and, in the case of a university press, the faculty advisory board, who make the final decision about publishing your book. Throughout this chapter, I also recommend methods for maintaining your dignity and, when necessary, asserting yourself.

FINE-TUNING YOUR PROPOSAL

If you have the opportunity to meet acquisitions editors and series editors in person, I encourage you to do so before submitting your proposal. The good impression that you make face to face will serve you well when you communicate screen to screen. But whether or not you're able to arrange such a meeting, a tidy, complete proposal prepared according to the publisher's guidelines will stand out.

Returning to your list of candidates, sort them in order of preference. Assuming that the acquisitions editor may request exclusive review, you may want to submit your proposal only to your top-choice publisher first. If you decide to submit to two or three publishers during the initial round, you should put a note in the cover letter, explaining that multiple editors have received the proposal but that you're willing to put the others on hold as soon as one editor begins peer review.

You already assembled your portfolio. Now it's time to fine-tune your proposal for individual publishers. Check the submission guidelines for each one. In addition to your CV and the table of contents with chapter summaries, the publisher may ask you to complete a form. Although the task seems elementary, please take the time to glue the relevant information into the spaces. Don't simply type "See CV." The form allows an acquisitions editor to assess your submission rapidly. The less time she spends searching for key information, the more time she can spend reading your proposal.

The standard form calls for two or three variations of the book description slanted toward different audiences. Think about which details to emphasize in your sales pitch for each group you want to reach. Some will care only about your topic, whereas others will want to hear your conclusions. Some will consider your expertise, experience, or accomplishments as important as your research, whereas others will take an interest in your methods and evidence.

The standard form also asks for a list of competing books. Because you've been paying attention to the newest literature in your field, you can simply draw on your citations to compile a list of four to six books published in the last three years. For each book, explain in one brief paragraph how yours differs in terms of scope, evidence, methods, and argument. On the surface, this question seems to ask you to justify publishing your own book. Rather than take the defensive, turn this opportunity into a strategic advantage. Think about all the ways the information you provide can influence the publisher's decision-making. If you include at least one of the publisher's books on the list, you demonstrate your familiarity with these titles. And that is a genteel form of flattery, is it not? As you survey the competition, you can suggest how to position your book in the market successfully. By comparing formats and prices for these books, for example, you subtly point to a competitive format and price for yours. You alert the acquisitions editor to authors who would make good reviewers or, conversely, those whose scholarship might compromise their ability to read your proposal with an open mind.

If you intend to take an active role in peer review, you should begin now by proposing a list of reviewers. Please put this list in a separate document so that the acquisitions editor can detach it before sending out the proposal package. No one knows better than you who has the qualifications to evaluate your work. Be aware, however, that the term *peer review* is somewhat of a misnomer. Colleagues at the same institution, graduate school classmates, and co-authors are barred on the grounds of a presumed conflict of interest. Former mentors and others with direct personal or professional relationships are also banned from the reviewer pool. Remember, you're not listing references as you would for a job application. Whom should you nominate? At a minimum, a reviewer should have a graduate degree, relevant experience, several publications, and a tenured position (if applicable in your field). For each name on your list, provide the person's contact information and area of expertise as well as the reason for recommending him or her. Yes, you should include a few individuals representing neighboring fields, but make clear why you consider them appropriate reviewers. Yes, you can make a short blacklist. All you need to say is "I respectfully request that the publisher does not invite the following individuals to review my proposal." You don't need to give an explanation; certainly you should not throw stones.

Returning to the generic cover letter that you drafted earlier, enter the date, the name and title of the acquiring editor, and the name and address of the press. In the final paragraph, mention the series for which you are submitting your proposal (if applicable) and list all the enclosed documents. Paste this letter into the body of an e-mail message and attach it as a PDF file. This is your last chance to make sure that you've assembled all the components of your portfolio: cover letter, CV, brief table of contents, table of contents with chapter summaries, introduction, writing samples, and submission form required by the publisher. For your own sake, please take the time to proofread each component. I recommend that, as a reminder to anyone tempted to co-opt your work, you add a watermark such as "under copyright" to each page. Indeed, as I explain in a future chapter, you own the copyright until you assign it to another party. Also, you would be wise to prevent accidental corruption by submitting the files as PDFs and to prevent confusion by choosing readily identifiable filenames (e.g., Herr.NuclearEnergy.ChapterSummaries. Dec2016).

For the subject line of your e-mail message, you want to use an equally identifiable label. May I offer an insider's tip? At a press, the publishing team refers to an individual book by the author's last name, followed by a backslash (/), followed by few keywords. For example, the team at Greenwood labeled this guide Herr / Writing and Publishing; my model proposal is Herr / Nuclear Energy. By adopting this structure in the subject line of your e-mail message—and using it every time you correspond with any member of the team—you not only make your messages easy to recognize, but equally to your advantage, you demonstrate your familiarity with publishing conventions.

After submitting your proposal, you may not receive a response for a few weeks. How long should you wait? Should you send the editor a reminder message? At what point should you telephone? I suggest that you wait two weeks then simply resend the e-mail containing your proposal. (You might wait three weeks over the holidays or during a national conference, such as the annual American Political Science Association meeting.) If, after another two weeks, you don't receive an acknowledgment, try phoning the acquisitions editor or her assistant. At that point, if you don't receive a response, move on to the next press on your list.

PEER REVIEW

An acquisitions editor may respond to your proposal in one of three ways: 1) a rejection without peer review; 2) a request that you revise the proposal; 3) a notice that she will begin peer review or, if peer review isn't required, she will proceed with the in-house approval process. In the first scenario, you might ask, in a nonconfrontational way, about the reason for rejection so that you can improve the proposal before submitting it elsewhere. In the second scenario, welcome the editor's suggestions to the extent that they align with your own vision for the book and propose a deadline for resubmitting the revised version. You might also request a phone appointment in order to gauge her interest level and learn how she intends to position your book among the press's other titles in your field.

In the third scenario, when the editor contacts you, ask her to lay out the entire review and in-house approval process, step by step, and estimate the time frame. I strongly recommend that you document your exchanges with the editor. Keep your e-mail correspondence and file it systematically so that you can easily locate particular messages. After every phone or in-person meeting, send the editor a summary of your notes and ask for confirmation that you accurately captured the conversation.

During your initial conversation with the editor, you want to inquire about the policy regarding exclusive review and multiple submissions. *Exclusive review* means that once the press begins reviewing your proposal, you must withdraw it from consideration elsewhere until the press makes a decision. You may be willing to give your top-choice press exclusive review, but you should inform the acquisitions editor that you're also exploring other presses. How long might peer review and the approval process take? Could she agree to a specified time period, say four months after the submission date? If the press does not make a decision by that deadline, you need to be free to submit your proposal elsewhere. By offering such a compromise, you balance compliance with self-assertiveness. And, of course, you should alert all the editors whom you approached that your proposal is under review at another press.

If this particular editor permits multiple submissions, you still want to notify the other editors. Some may respond by starting peer review immediately; others may put your proposal on hold until you receive a decision from the first publisher. Whatever the case, you should keep everyone informed about the status of your proposal. Remember that editors working in your field probably know one another, and they will likely compete for the same reviewers. You don't want an editor to be told by her colleagues or reviewers that your project is under consideration at another press.

During that initial conversation, you should also inquire if review of the proposal suffices for publication or if there will be additional layers of review. The university presses for which I acquired, for example, could offer an advance contract when a proposal received laudatory reviews but then required peer review and faculty advisory board approval of the completed manuscript before publication. The more you know, the better you will navigate the process.

If given the opportunity, you want to write two or three of your own questions for peer reviewers. The press sends a set of standard questions regarding the quality of your research, the force of your argument, and the originality of your conclusions; yet a few additional questions from you demonstrate receptiveness to criticism. At the same time, you solicit advice on specific concerns, such as gathering additional data, graphing results, reorganizing a chapter, or including a map.

In the science, technology, engineering and mathematics (STEM) fields, journals conduct double-blind reviews, meaning that the editor reveals neither the author's identity to the reviewer nor the reviewer's identity to the author. At the other extreme, a few journals are experimenting with "open review," also known as "community review," in which self-nominated or invited individuals publicly comment on a manuscript posted online and all participants identify themselves. Taking the middle way, the majority of book publishers conduct single-blind reviews, meaning that reviewers remain anonymous but they are given the author's name and CV.

Generally, an editor obtains two peer reviews. For a multidisciplinary project (e.g., one that touches on psychology, public health, and regulatory science), she may collect more reviews in order to ensure coverage of all the relevant fields.[1] For a classroom book, she may layer peer review with an informal market survey and send the proposal to six or eight instructors representing different types of public and private institutions. Recruiting reviewers and then pressing them to meet a deadline requires unflagging persistence coupled with gentle tact. For their part, reviewers deal with professional and personal obligations while pursuing their own research. Moreover, those high-ranking scholars most qualified to evaluate your book proposal receive scores of requests to review journal articles, grant proposals, and book manuscripts every year. Reviewing even a short, well-put-together book proposal consumes at least an entire workday.

So, you ask, if peer review causes authors so much anxiety and costs publishers and reviewers so much time and labor, why bother? Because it makes better books. Reviewers note weaknesses in your argument, suggest new literature, detect factual errors, point you to additional data, or recommend further analysis. Yes, they ask you to do more work, but only to avoid embarrassing yourself. Reviewers hold key positions in your field and possibly neighboring fields; thus peer review extends your professional network and alerts future audiences about your book. The very fact of peer review stamps your book with a quality assurance seal, giving it credibility with discerning readers. With the right attitude, you'll even find peer review enjoyable. Reviewers read your proposal attentively, reliving your research and engaging with your ideas, with the aim of helping you. What could be more gratifying?

For a book proposal, an editor gives reviewers four to six weeks to return their comments. If the deadline passes and you haven't heard from the editor, send her a query. Here again, expect delays over the holidays and during major conferences; but don't hesitate to resend your query two or three times over the next two weeks. Most likely the editor is not ignoring you but rather holding out in hope of receiving the reviews. If you haven't received them within four weeks after the reviewers' deadline, it's time to pick up the phone. Listen to the editor's explanation of the delay, but politely inform her that if you don't have the reviews by a specified date, you must submit your proposal to another press. Then do exactly that.[2]

READING THE REVIEWS

Some reviews cheer with praise; some chill with criticism; most shift back and forth like the temperature on a breezy autumn day. Here I present a few hypothetical scenarios and discuss how to respond to each. First, in the unlikely event that both reviewers discover flaws in your research or conclusions, the acquisitions editor will reject your proposal. I say "unlikely" because an experienced editor usually detects such flaws without the need for review. It is equally unlikely, however, that both reviewers will give you an ovation.

Such ebullience might cause the editor to wonder how attentively the reviewers read your proposal or how an undisclosed conflict of interest influenced them. But if she basically concurs with their assessment, she will take your proposal to the next level for approval. One other fairly uncommon scenario involves split reviews: Reviewer A sees only the virtues of your work, whereas Reviewer B sees only its faults. In this case, the editor could recruit a third review to break the tie. Alternatively, if she has a graduate degree in your field or other credentials, she herself may make the call, according to her confidence in your book's potential.

Most likely, each review will combine praise, criticism, and observation. Along with her own comments on your proposal, the editor may give you her interpretation of the reviews or highlight points that she wants you to address in your response. Whether or not she provides this guidance, you can make the reviews manageable by separating the comments into categories. I find the following method intellectually effective because it helps you to organize the feedback and emotionally satisfying because it lets you take scissors to the reviews—literally. First, print two single-sided copies of each review. On one copy of Review A, use a colored pen to mark the compliments. On the other, use a different color to mark the criticisms. Cut out each compliment and criticism, then sort the pieces into these piles:

* Research and Analysis—nature, scope, and validity of your data; appropriateness and accuracy of your analysis
* Argument—how convincingly your research and analysis support your conclusions
* Presentation—the overall structure of your book; the organization of individual chapters; the quality of your writing; the intelligibility of your graphs, tables, and illustrations

Put the remaining snippets in another pile for observations about your topic or the field in general. Don't be surprised when a reviewer tells personal anecdotes or gives a discourse seemingly unrelated to your proposal. You can put aside such tangential commentary. Repeat these steps for Review B, this time using pens of two other colors (that is, use one color for praise and one color for criticism for each review).

Allow yourself a few days to process the piles one by one, revisiting your proposal and considering the validity of each complement and criticism, with a view to writing your response letter and revising the plans for your book. Do you need to gather more data or run additional analyses? Is your argument altogether too weighty for the evidence you can assemble? Or does it wobble because you haven't fortified the connections between your research and your conclusions? Do the reviewers demand reorganization, or do you simply need to clarify the logic underlying the current structure of individual chapters and the book as a whole? Do the visuals convey information clearly and intelligibly? What alternatives could you try? Maybe you should explain more in prose?

A reviewer may complain about substance when actually the problem lies with the presentation or vice versa. Despite his inability to articulate precisely what troubles him, you must attempt to discern the cause and apply the remedy. Keep in mind the possibility that he read your proposal too quickly or didn't comprehend it fully. Also remember that he lacks your deep expertise. You are the world's number-one expert on your topic. Yet, as an author, you shoulder the task of explaining your work and its import to those of us who know much less than you do. Be patient.

YOUR RESPONSE LETTER

Now that you've made the acquisitions editor your ally, it's time to launch the second phase of your charm offensive targeting the in-house editorial committee and, if you are working with a university press, the faculty advisory board. You must persuade them, first, that you will work collaboratively with the publishing team and, second, that together you will produce a book worthy of the press imprint. Of course, they will read your words but, even more, they will listen for your tone. For them, the ideal author is someone who balances confidence in his expertise with the humility to welcome well-meaning criticism. As you craft your response, demonstrate that you make deliberate choices about your book grounded in the logic of your research and, at the same time, you recognize the value of the peer review.[3] You are equally a self-assured expert and an amiable colleague.

Unless your editor requires a point-by-point response to each review, I suggest that you limit your letter to one and a half pages, two pages at most. Begin by thanking Reviewer A and Reviewer B for their evaluations. Mention how gratifying you find their praise; note the particular features of your project that they lauded. Then discuss their chief criticisms and how you intend to address these as you write the book. You may explain why you made certain choices, but take care to avoid a defensive tone. When you reject a reviewer's suggestion, provide a convincing reason such as unobtainable data or the intentionally defined scope of your project. If the reviewers give you contradictory advice, explain which path you're choosing and why. If they raise concerns about your writing or presentation, declare your eagerness to work with your editor and the publishing team. Close your letter by thanking your editor for collecting the reviews and championing your project.

FINAL LEVELS OF APPROVAL

Your editor is indeed your champion, and you'll be glad you made the extra effort to cultivate a collegial partnership when she takes your proposal to the next level of the hierarchy. The in-house approval committee typically consists of representatives from the acquisitions department and the marketing department; the press director may also participate. About a week before the committee meets, your editor distributes a packet including a description of your project, a summary of your qualifications, selections from your proposal, the complete reviews, your response to the reviews, a list of selling points, a survey of the competition, a list of comparable books recently published by the press, and a cost-and-income projection. Every press has different specifications for this packet, but these are the most commonly required components. I'm giving you this complete list both so that you can assist in compiling the information, much of which you already supplied in your proposal and on the submission form, and so that you can see how thoroughly your editor prepares to advocate for your book.

During the committee meeting, your editor gives a short presentation outlining your book's merits and market potential. After making a case for the book itself, she discusses how it fits the publishing strategy for your field. Finally, she answers the committee's questions, ranging from her choice of reviewers to the cost of illustrations, from the nuances of your analyses to the relevant book prizes. Throughout the meeting, she strives to convince the committee that, together, you and the publishing team will produce a financially viable book that benefits readers and enhances the reputation of the press.

The committee could reject your proposal outright. More likely, if the members have concerns, they will ask your editor to gather more information or discuss details with you and return your proposal to a future meeting. Ideally, they will agree to offer you an advance contract. Before you sign it, however, you want to ask about the remaining requirements for publication because "advance contract" has different meanings for different publishers. At a university press, for instance, an "advance contract" requires that your finished manuscript undergo another round of peer review and receive the endorsement of the faculty advisory board.

Let me pause for a moment to describe the faculty advisory board's role. Composed of tenured faculty representing the various fields in which the press publishes, a board oversees the scholarly functions—as opposed to the business functions—of the press to ensure that every publication is worthy of the university's name. A week or two prior to the board meeting, your editor distributes an information packet similar to the one that she prepared for the in-house approval committee; during the meeting, she gives a similar presentation. In my experience, board members steer wide to avoid books that donors, taxpayers, or alumni may perceive as scandalous, polemical, or biased in any way. Only impartial research and scholarship, conducted according to the highest professional standards, is acceptable. Although board members will not hesitate to offer their own comments on your manuscript, primarily they will scrutinize the reviewers' qualifications, the quality of the reviews, and the cogency of your response. The board may decide to approve your project, stipulate additional revisions, or require another review. Whereas the board has the authority to cancel a publishing contract, it rarely issues an outright rejection because an experienced editor and her colleagues on the in-house approval committee cull projects with poor prospects long before they reach the board. Speaking again from my experience, I can attest that a board neither rubberstamps nor mindlessly discards a project. You should expect a fair hearing. You should also feel honored that so many fellow researchers and scholars, experts in their own right, are giving your work attention.

TAKING THE NEXT STEP

Hopefully, your proposal sails smoothly through the submission and review process. However, if you receive a rejection letter or if you become dissatisfied with the way the acquisitions editor is handling your proposal, take the initiative. Move on to the next publisher on your list. First, consider revising your proposal. Did the editor give you any helpful tips? Did the reviewers make sensible recommendations? If so, take advantage of them. Second, revise your cover letter, complete a new submission form, and tailor your proposal to meet the next publisher's requirements. Please, for your own sake, don't simply resend the same e-mail with the same proposal. You will lose points instantly if you send Routledge a proposal addressed to Cambridge University Press. No acquisitions editor wants to discover she is your second choice!

Don't give up until you've tried all the candidates on your short list. But then, if you are still struggling to find a publisher, you might want to reconsider your plans. Maybe you need to develop your project or your professional credentials further before you can attract a publisher. Maybe a book is not the best publishing option for this particular project after all. Nonetheless, by creating a book proposal and submitting it to various publishers, you gained a deeper understanding of your project, your field, and your audiences as well as an introduction to book publishing. Congratulate yourself on your growing expertise.

If a publisher offers you a contract, celebrate! You have a partner—indeed, an entire team of publishing professionals—ready to assist you in producing and marketing your book. But before you sign your contract, you want to understand your rights and responsibilities. In the next chapter, I explain the publishing contract and empower you to negotiate with your prospective partner.

NOTES

1. The Association of American University Presses published a guide to best practices for peer review. Although this guide is explicitly written for editors, an understanding of the general rules of the game will benefit you. AAUP Acquisitions Editorial Committee, *Best Practices in Peer Review* (New York: Association of American University Presses, 2016), accessed 28 December 2016, http://www.aaupnet.org/policy-areas/peer-review.
2. Rachel Toor likewise assures authors that they need not become victims of an unconscionable delay. If you do not receive peer reviews within six months, she suggests that you submit your proposal to another publisher. Rachel Toor, "Held Hostage at a University Press," *The Chronicle of Higher Education*, 9 December 2013, accessed 28 December 2016, http://www.chronicle.com/article/Held-Hostage-at-a-University/143523?cid=megamenu.
3. Wayne C. Booth et al. suggest additional ways in which to frame your formal response letter and to counter anticipated criticism in your text so as to balance acceptance of criticism with confidence in your work. Wayne C. Booth, et al., *The Craft of Research*, 4th ed. (Chicago: University of Chicago Press, 2016), 141–54.

CHAPTER 6 _____

Understanding Your Publishing Contract: The Author–Publisher Partnership

A publishing contract is a business contract outlining the agreement between two partners—the publisher and the author—to produce a well-written, attractively designed book with strong market potential. Each partner contributes a form of capital. The author offers the content and a select package of rights for its use, and the publisher supplies the operating facilities, raw materials, and skilled labor. In this chapter, I discuss the key contract provisions that lay out the complementary roles as well as the respective rights and responsibilities of the two partners.

A MUTUAL COMMITMENT

The publishing contract opens by introducing the two members of your partnership (the Author and the Publisher) and the object of the commitment (the Work, which is your book) then proceeds to enumerate the roles of the partners throughout the book's life cycle.[1] The contract is divided into numbered sections called *clauses*, which are further divided into subclauses designated by lowercase letters. For example, §14(b) indicates clause 14, subclause (b). This fine-grained structure allows the contract to specify each right and each responsibility concerning the Work. If you're writing the book with one or more colleagues, you and your co-authors collectively constitute "the Author." Consequently, you want to draw up a formal agreement among yourselves regarding the order in which your names will appear on the title page, the division of royalties, and the allocation of Author rights and responsibilities. Add this agreement to the Publisher's contract as an addendum.

COPYRIGHT AND OTHER RIGHTS

Typically, the first clause in a publishing contract addresses a collection of rights packaged under the label *copyright*. According to the *U.S. Copyright Act*, the Author automatically owns these rights as soon as the Work becomes "fixed in any tangible medium of expression" (§ 102).[2] The moment you write, type, or record your words in any way, you own the rights pertaining to them. If you have co-authors, all of you equally share ownership (§ 201(a)). There are a few exceptions. If you conducted the research or wrote the manuscript for a commissioned project (i.e., a work-for-hire), you may have granted copyright to the sponsor (§ 201(b)). Likewise, if the research or writing is part of your job duties, copyright may belong to your employer. Check the institutional or corporate policy to ensure that you own copyright and that you have the authority to negotiate with the Publisher.

You don't need to register with the Copyright Office in order to claim copyright for anything you produce; and for conference papers, lecture notes, and similar unpublished writing, formal registration has little benefit. But for published writing, registration makes copyright ownership easier to prove and, thus, legal protections easier to enforce. The Publisher takes responsibility for copyright registration, either in its own name or in the Author's name. As an author and as an acquisitions editor, I've heard both sides of the debate over which party should own copyright. As far as I can tell, the debate springs from the suspicion that, at some future time, someone somewhere will invent some new means of making a profit from the book and whoever holds copyright will automatically control the right to implement these means. But no matter which side of the table I'm sitting on, the debate seems trivial because, regardless of who owns copyright, the Publisher still requires certain rights from the copyright package in order to fulfill its immediate responsibilities. Moreover, based on my experience with the advent of e-books and myriad forms of digital distribution, I predict that when this new invention comes on the scene, the Publisher and the Author will profit most if they collaborate.

The law enumerates the rights covering all classes of scholarly and artistic creations eligible for copyright protection. Three are especially pertinent to a book publishing contract: reproduction, derivative works, and distribution (§ 106). In other words, the law grants the copyright owner the rights 1) to make multiple copies of the Work; 2) to adapt the Work to create new products; and 3) to sell, rent, loan, or give away copies of the Work. The publishing contract contains not only the main copyright clause, which consists of six or eight subclauses, but also several other clauses regarding various aspects of copying, adapting, and distributing the Work. By breaking apart the copyright package in this way, you and the Publisher clarify precisely what may be done with your book. But the contract speaks in abstract terms, and you may be asking what these rights mean in practice.

What does it mean to reproduce the Work? First, even for a straightforward copy of the text, there are a multitude of format options: hardcover and paperback print editions, half a dozen e-book varieties, and numerous reading apps for mobile devices. There are an equal number of format options for audiobooks: CDs, streaming, digital downloads. The contract transfers to the Publisher the right to make copies of the Work and specifies the permitted formats. Because technology changes so rapidly, your contract most likely contains a phrase such as "formats now known and yet to be invented." This provision gives the Publisher flexibility to pursue opportunities for strategically increasing your book's availability.

What does it mean to create derivative works? The standard publishing contract uses the term *subsidiary rights*. For nonfiction books, the most common derivative works are translations. Repackaging all or part of the content is another way of producing derivative works. For example, the Publisher might include one of your chapters in an anthology on key social issues. Alternatively, the Publisher may sell to a popular magazine the right to publish an excerpt from your book. In the publishing industry, prepublication rights are known as *first serial rights* (i.e., the excerpt appears in print before the publication of your book), whereas postpublication rights are known as *second serial rights* (i.e., the excerpt appears in print after your book's publication). The contract may also mention film, dramatization, or other creative adaptations based on the Work. Although a documentary film maker might find inspiration in a biography, scholarly and professional books are rarely candidates for box office films or Broadway musicals.

What does it mean to distribute the Work? The contract identifies the territory in which the Publisher may act as an agent for the Work—typically the world. It also specifies the ways in which copies of the Work may be made available. Once readers obtained books primarily by purchasing them or borrowing them from a library. But now, thanks to new formats and the new distribution systems that support them, publishers not only sell books but also rent them and make bundles of selected books available by subscription. Publishers also donate books or use them as promotional giveaways.

The Publisher requires these rights in order to operate effectively and strategically in today's complex, constantly changing business world. In addition, the Publisher serves as a broker for these individual rights and, with each transaction, both you and the Publisher benefit. The Publisher's Rights and Permissions Department has the expertise, experience, and reach to market translation rights and other subsidiary rights to publishers and media outlets around the world. When the Publisher sells particular rights or leases them for a period of time, you receive a percentage of the profit.

You should not hesitate to inquire how the Publisher intends to promote your book as well as the package of rights associated with it. By asking these questions, you learn more about the Publisher's commitment to your book, how it fits into the Publisher's larger business plan, and to what extent the marketing team welcomes your suggestions. You get a better sense of how much the Publisher values the partnership and how well it will function. That said, you want to ensure that when you transfer the package of rights to the Publisher, you can use your book for select professional purposes. Do you have the Publisher's permission to post a segment of your book on your own website and deposit it in an institutional or disciplinary open access repository? May you use segments of your book for teaching or convert a chapter into a journal article?

THE PUBLISHER'S COMMITMENT

The contract spells out the Publisher's responsibilities, which include payments to the Author, publishing services, and marketing efforts.

Royalties

As I mentioned earlier, each publishing partner contributes a form of capital to the business. The Author supplies the Work and a package of rights associated with it; the Publisher provides the facilities, materials, and labor. Significantly, the Publisher makes the up-front investment and assumes the financial risk. As the Author, of course, you invest a great deal in research and writing. But even if the book is not financially successful, you will not be out-of-pocket for the production and marketing expenses. On the other hand, as soon as the Publisher begins to recover its investment, you receive a portion of the revenue. It may be useful to think of royalties as a type of profit-sharing.

Just as the early clauses in the contract specify the formats in which and the means by which the Publisher may distribute your book, later clauses specify the royalty payments for each of these formats and means as well as for the sale or lease of rights (e.g., translation, film adaptations). Royalty clauses are extraordinarily complex, but they become intelligible once you understand a few key terms.

Revenue or *income* refers to the total amount of money received by the Publisher for a transaction, such as the sale of Japanese translation rights. For this kind of one-time transaction, the Author receives a fixed percentage of the revenue in a single payment. For recurring transactions, the Author typically receives graduated royalties: the percentage of revenue increases as the number of transactions, and thus the margin of profit, increases. For the sake of illustration, I present a few scenarios.

The Publisher assigns your book a recommended retail price, called the *list price*. But the Publisher sells books to retailers at a discount ranging between 30 percent and 45 percent of the list price. If the list price for the hardcover edition is $49.95 and the discount is 30 percent, the Publisher receives net revenue or net sales of $34.97 for each copy sold to retailers. If the contract grants you 5 percent of the list price for the hardcover edition—which would be quite exceptional—you receive $2.50 for each copy sold. Authors of professional and scholarly books generally receive royalties based on net revenue. Therefore, if the contract grants you 5 percent net revenue for the hardcover edition, you receive $1.75 (that is, $34.97 \times 0.05 = 1.75).

Continuing with this example, let's look at a typical arrangement for graduated royalties. For the hardcover edition, the Publisher offers you 2 percent net revenue for the first 500 copies sold, 5 percent net revenue for 501 to 5,000 copies sold, and 7 percent net revenue for all copies sold thereafter. Assuming that the list price and the discount for retailers remain stable throughout the book's life, you receive $0.70 for each of the first 500 copies sold, $1.75 for each of the next 501 to 5,000 copies, and $2.45 for each of the rest of the hardcover copies.

Such a royalty schedule makes sense when you consider that nearly all the costs of publishing a book are up-front costs. Before selling a single copy of the book, regardless of its format, the Publisher must pay acquiring editors, copyeditors, production managers, graphic designers, printers, web designers, marketers, warehouse workers, and information technology professionals. The Publisher prices a book so as to recover this investment from the sale of a certain number of copies. Beyond this point, because ongoing expenses for marketing, warehousing, and distribution are minimal, the unit cost drops, resulting in a greater profit margin. For the same reason, the Publisher may ask you to waive royalties on the first few hundred copies. Then, when the revenue from your book has repaid the initial publication costs, you will begin to receive royalties.

The contract stipulates how often you will receive a royalty statement, with a summary of sales and the income owed to you, and a royalty check. Although you should receive a statement every 6 or 12 months, you may not receive a check with each statement. Because of the cost of processing a check, most publishers will not write one unless the royalties for the statement period meet a minimum, such as $50. Only when the royalties accumulate to the specified minimum, or after a given period of time, will you receive a check.

A royalty advance, rarely offered for professional and scholarly books, is a payment that the Author receives before the first sale—that is, in advance of standard royalty payments. Essentially, it is a loan. Say the Publisher agrees to pay you a royalty advance of $3,000 in two installments: $1,500 when you sign the contract and $1,500 when you deliver the finished manuscript for publication. Although you will receive regular statements, you won't receive a check until your book has earned back the $3,000 in royalties that the Publisher already paid you. A royalty advance serves as a demonstration of each party's commitment to the partnership. If for some reason you fail to deliver the finished manuscript, you are obligated to return the $1,500 received upon signature of the contract. However, if the book does not sell as well as anticipated and the royalties never reach $3,000, it is unlikely that the Publisher will demand that you return the advance.

Publishing Services

The contract clauses regarding the publishing process specify which partner takes responsibility for each step. In the sciences and the quantitative social sciences, it is not unusual for the Author to submit a PDF with all the pages, including the index, laid out exactly as they will appear in the published book. (The contract may refer to this PDF as *camera-ready copy*, a term surviving from the predigital era.) In this case, the Publisher gives the pages a light overall copyediting or limits copyediting to the title page and the table of contents. If the contract stipulates a printer-ready PDF and you lack the necessary editing and design skills, ask your acquisitions editor to recommend a freelancer whom you can hire. In the humanities and the qualitative social sciences, the Author typically submits a manuscript.[3] The Publisher provides full copyediting and design services, although the Author creates the index before the book goes to press. You want to ensure that your contract spells out precisely what the Publisher will do for your book. If the wording seems vague, ask for clarification.

At the same time, do not be alarmed when the contract says that you will be charged for changes to the page proofs. This clause is intended to prevent the Author from rewriting the book the moment before it's due at the printer. As you will discover when you see your own page proofs and realize that you have one last chance to perfect your book, the urge to revise every sentence is hard to resist. If the Publisher intends to charge you for any type of copyediting or design, however, you may be dealing with a so-called vanity press. Look again at the previous chapter on investigating your publishing options and consider carefully whether or not this press can help you achieve your goals for your book and, ultimately, for your career.

Publication Schedule

Contract clauses pertaining to the publication schedule are less explicit than those pertaining to rights and royalties. The contract should guarantee publication of the Work within "a reasonable amount of time" or within a specified period after delivery of the final manuscript. Given that many authors underestimate the time needed to write a book, the Publisher cannot make firm plans too far into the future. You might inquire about the anticipated timetable from manuscript delivery to official publication. For professional and scholarly books, the industry standard is 8 to 18 months. Does the book's sales potential depend on timing? Is there an upcoming event or anniversary that will launch your topic into the news? By alerting the Publisher to these opportunities, you can help determine the ideal publication date. But keep in mind that production cannot begin until you finish the manuscript.

Marketing and Distribution

Before choosing where to submit your book proposal, you did a good deal of research on publishers. Now that you are about to partner with this one, look at your notes again and work in reverse. Where and how did you find this press? At a conference? On the web? On your bookshelves? Did a sales rep contact you or your colleagues? Did you spot an ad in a conference program or in a professional journal? Was there an ad on a website? Did you receive an e-mail announcement about new books? Will the Publisher use the same channels for promoting your book? Contract clauses pertaining to marketing and distribution tend to be less than explicit, so don't hesitate to ask your acquisitions editor for a draft of the marketing plan.

THE AUTHOR'S COMMITMENT

By signing the contract, you acknowledge your legal accountability for your book's content and accept a number of obligations. First, the warranty and indemnification clause declares that you, the Author, have the authority to sign the publishing contract, that the Work is original and factual, and that the Work contains nothing libelous. Further, this clause states that you do not infringe on the copyright or intellectual property rights of others. In practice, this means that you must determine whether or not the ways in which you use quotations, images, and other material drawn from various sources meet the requirements for fair use. If not, you must obtain permission from the rights owners. You may consult your acquisitions editor; but ultimately you are responsible for citing sources properly, applying fair use appropriately, and paying permission fees. (At this point, you might want to take a quick look at the chapter on different types of evidence where I discuss fair use and permissions.) Lastly, the warranty and indemnification clause outlines the process for responding to an accusation that the Author has trespassed on the rights of other parties.

Manuscript Due Date and Specifications

Without a doubt, the most-often-violated clause is the manuscript submission clause—and this is the biggest cause for tension between the two partners. The Author understandably views the Work as unique, rigorous, and artistic. The creative process must not be rushed, and the Work itself must not be constrained. Despite an appreciation for the Author's expertise, the Publisher views the Work as a commercial product. It certainly is unique, yet it is one among many products from many suppliers (i.e., Authors). Financial success depends on a reliable input of manuscripts with predictable production costs and publication schedules. Ideally, these two perspectives complement one another so both partners want to take a realistic look at the manuscript submission clause.

Before committing to a due date, ask yourself a few questions. What are my current professional responsibilities? Will I have to take on additional work when a colleague retires or goes on medical leave? Am I about to be promoted to a time-consuming administrative position? What about my current personal responsibilities? Am I planning a wedding, a pregnancy, a move, a trip to New Zealand? Are my family members healthy? Given these responsibilities, how many hours each week can I devote to writing? At this rate, how many weeks do I need to finish the manuscript? When you arrive at an answer, add three months. Then, with a realistic estimate of your own timeline, you can negotiate the due date. The Publisher will appreciate your candor and, ultimately, reward your reliability.

The Publisher will likewise appreciate your candor and reliability with regard to the manuscript specifications. For the sake of financial planning, the Publisher calculates the production costs for your book based on the word count and the number of tables, graphs, and images. Can you agree to these specifications? Estimating the word count can be difficult. For example, say you've written one of the eight chapters. Run a word count on this chapter—don't forget to include the footnotes or endnotes—and multiply by eight. The result is the estimate for the chapters. Together, the introduction and the conclusion usually equal one chapter, so add another chapter's worth of words to your estimate. In a typical nonfiction book, the list of references contains approximately as many words as one chapter. Here's the equation for calculating the word count:

chapters (including footnotes or endnotes)

+

introduction and conclusion

+

list of references

Is your estimate within 5,000 words of the Publisher's? If not, you want to discuss the desired length of your book with your acquisitions editor.

Likewise, if your estimate for the number of tables, graphs, figures, or other images does not match the Publisher's, you want to talk about how these elements enhance the book and how to balance the benefits against the costs. Yes, spectacular photographs and intriguing drawings attract readers. Yes, the cost of printing images has dropped and the cost of reproducing images in digital books is negligible. Yet permission to use images in various formats can be extraordinarily costly, and you may be responsible for this expense. Now is a good time to consider alternatives.

Finally, if you use foreign-language characters or mathematical symbols, you want to make sure that the Publisher has accounted for them in the cost estimate. Is there a limit to how many you may include? Do they need to be formatted in a particular way?

Formatting the Manuscript

Along with the contract, you should receive formatting instructions, detailing how to prepare the manuscript before you submit it for publication. Although these instructions are not officially incorporated into the contract, failure to follow them will delay publication. Does the Publisher require a fully designed PDF (aka, camera-ready copy) or a manuscript prepared using conventional word-processing software? The submission guidelines specify the page layout: line spacing; margin size; font size; numbering scheme for pages, notes, tables, graphs, and images. They also indicate the recommended citation style, the Publisher's preference for footnotes or endnotes, and the placement of the nontextual elements. By implementing the guidelines as you write your manuscript, you will spare yourself weeks—literally weeks—of work.

Publishing Process

If the Publisher requests a fully designed PDF, you have minimal responsibilities during the publishing process. Your team copyedits the front matter, creates your book cover, and handles the printing; you need only review the design and the page proofs (i.e., the last version of your book before it's sent to the printer). If you submit a manuscript, the Publisher will hire a copyeditor to correct your text and a designer to lay out the pages and create your book cover. You review their work with the assistance of the production manager overseeing the entire process. The contract indicates how much time you have to review the copyedited manuscript and the page proofs.

Most publishers delegate the index to the Author. Because you know your topic better than anyone, you can anticipate the topics of interest to your readers and ensure that all the essential terms appear in the index. Thankfully, unless you submit a fully designed PDF, you won't create the index until you receive the page proofs. And you can hire a professional indexer to do the hard work; all you need to do is assist with compiling the terms. (If you want to create the index yourself, you'll find suggestions in the upcoming chapter on publishing and marketing.) Here again, if the contract fails to spell out your obligations, ask for clarification. If the obligations seem unreasonable, ask what's negotiable.

ADDITIONAL CONSIDERATIONS

Before signing the contract, there are a few other issues you want to discuss with your acquisitions editor. First, especially if you are signing an advance contract, it should spell out the requirements for publication. Will your manuscript undergo additional rounds of peer review or editorial review? Who will make the final decision about whether or not your manuscript is ready for publication? If the Publisher deems it unacceptable, will you have the chance to revise it? If the mandated revisions run contrary to your own vision for your book, can you be released from the contract?

Second, what is the Publisher's plan for the initial publication? Will your book appear in hardcover, paperback, and digital versions simultaneously? Will the paperback follow the hardcover within a certain number of months? Please note that the Publisher may not commit to a paperback in the contract but prefer instead to make this decision based on sales of the hardcover. Assuming your book sells well, does the Publisher expect to issue new editions? To what extent will you participate in these updates?

During this conversation, inquire about open access options as well. How much of your book will the Publisher make available for reading online without charge on its website or in a repository? How much of your book may you post yourself? What provisions can the press offer if you received a research or publication grant stipulating open access?

How many gratis copies of your book will you receive? Are you eligible for a discount if you purchase additional copies for personal use, such as gifts and tenure portfolios?

Will the Publisher handle all the promotional copies, or are you expected to purchase some for blurbists and book review editors?

In addition to a clause regarding future editions of your current book, most likely the contract contains a clause, known as the *option clause*, regarding your next book. When you're ready to circulate your new proposal, this clause obligates you to send it first to the Publisher, who has the right to exclusive review for a specified time, typically 90 days. At the end of that period, if the Publisher hasn't made an acceptable offer, you are free to send your proposal elsewhere. In my experience, this clause is negotiable, yet I encourage you to think of it as a mutual courtesy. The Publisher expresses interest in your next book even before your current book earns a penny while you, the Author, demonstrate confidence in the Publisher even before you see your current book in print.

Finally, if any concerns arose during your thorough investigation of prospective publishers, now is the time to discuss them with your acquisitions editor. Maybe another press's cover designs caught your eye; perhaps a particular marketing technique snagged your attention. On the other hand, you might've heard tales about long publication delays or lackadaisical copyediting. Although the publishing partnership involves continual give-and-take, I encourage you to use the contract negotiation process as an opportunity for an open conversation about the limits of compromise.

A DEVELOPING PARTNERSHIP

Now that you understand the terms of the Author–Publisher partnership, you're ready to sign the contract. In closing, however, I want to point out that it is only a paper agreement. The expectations for your partnership and the understanding of each partner's role develop through many conversations over a long period. You may encounter unforeseen circumstances during the life of your book; but by initiating frank conversations about the contract, you lay the foundation for future negotiations.

Let me also point out that the relationships that you cultivate with your entire publishing team, more than the contract formalities, affect your satisfaction with both the process and the product. Without mutual trust and respect, you have to expect mutual misery. Treat your teammates as professionals and gently insist that they treat you the same way. This relation-ship is key to a successful partnership.

NOTES

1. Following the conventions of the publishing contract, I capitalize the words *Author* and *Publisher* when I refer to the parties to a contract; I capitalize *Work* when I refer to the object of a contract. Although I make every effort to ensure the accuracy of the information provided in this chapter, it should not be construed as legal advice.
2. *U.S. Copyright Act, 17 U.S.C.* §102, accessed 28 December 2016, http://www.copyright.gov/title17/. In the following pages, I refer to specific clauses of this law in parentheses.
3. Technically, *manuscript* refers to a handwritten document. The accurate term for a typewritten document is *typescript.* Generally, however, these two terms are used interchangeably.

CHAPTER 7 ————————————

Drafting Your Manuscript: Performance and Presentation

For the first assignment in English Composition, my instructor asked the class to write a one-page essay using an analogy to describe the writing process. My fellow undergraduates compared writing to backpacking in the Adirondacks, bicycling in the Alleghenies, competing in the Ironman Triathlon, and building a log cabin, from cutting the timber to laying the roof, with hand tools. This clever assignment allowed us to complain and, simultaneously, forced us to articulate our resentment toward the required class. Every one of our analogies involved a physically demanding challenge. Writing, we agreed, is hard work! It's painful. It's risky. It's also intense and exhilarating. It transports the writer to a dimension outside of normal time and space. Even now, three decades and over a million pages later, I feel the same way about it. But now I know that although writing doesn't get easier, practice makes an author skillful, better equipped to handle the relentless challenges and able to reach the euphoric heights.

In this chapter, I propose strategies for writing well. You're the expert on your subject; I don't presume to tell you *what* to write. Besides, you already designed your book when you developed your proposal. Nor do I presume to lecture you on the elements of style. Experience as a reader and as a writer has taught me that there are just two essential components of good writing: presentation and performance. How you arrange the content determines readers' ability to assimilate the vital, complex information packed into your book. At the same time, how you deliver that information—your word choice, tone, and tempo—affects their attention level. A perspicacious author measures success by the readers' comprehension and engagement.

In the first pages of this chapter, I also touch briefly on the challenge of persistence. You don't need me to tell you how to manage your time or nourish your spirit. If you seek such guidance, you'll find plenty of specialists ready to dispense their wisdom. I simply encourage you to trust yourself.

THE URGE TO WRITE

With your publishing contract in hand, you can concentrate on writing the manuscript. Here's your opportunity for creativity, not just in the product but in the process. Break all the rules—even your own. Nothing says you must write the first chapter first or the last chapter last. You don't even need to start a chapter at the beginning. Start with something easy. What seems difficult now will become easier later because you'll have gained experience as a writer and, perhaps, circumstances will provide unexpected inspiration. Maybe you're working on a lecture or a conference paper related to Chapter 3. Maybe a rival just published an article that you must refute in Chapter 4. Or maybe legislators' stupidity in the face of a national crisis compels you to explain the practical applications of your analysis in your conclusion. What do you, the expert, have to say to the world right now? Write that, right now.

Whenever you feel the urge, write. Aspiring novelists, assuming there's a magic method, repeatedly ask famous authors about their habits. Do they write in the morning or the evening? Do they write every day? How many hours do they work at a time? Sorry to say, attempting to appease another writer's muse with another writer's sacraments won't help you. Find your own muse and create your own rituals to draw her favor. In other words, observe your natural cycle and then take advantage of it.

I'm not saying this to cajole you. I'm passing along valid conclusions derived from recent research on the work habits of successful academic writers. For decades, graduate students and junior faculty were told that writing every day for a specified amount of time is the best guarantee of creative productivity. Putting this conventional wisdom to the test, Helen Sword, herself an industrious scholar who offers writing workshops for fellow scholars around the world, interviewed 100 well-published academics and collected questionnaires from more than 1,200 others in the rank-and-file. "Write every day," she concludes, is not the only way. In fact, Sword finds that "roughly seven out of eight academics surveyed do not write every day; daily writing turns out to be neither a reliable marker nor a clear predictor of overall academic success." Interviewees report idiosyncratic practices: writing in the morning, the afternoon, the evening, or the middle of the night. Some unplug e-mail; some check it constantly as they write. Individuals write at work, at home, in exotic places, at nearby getaways, on the train, or on the couch. Whereas some prefer to write in small, daily sessions, many go on extended writing binges.[1] Sword's research proves that you don't need to squeeze into a one-size-fits-all writing practice.

If you need further assurance, read a few of the interviews with noted scholars published by Rachel Toor in her *Chronicle of Higher Education* column "Scholars Talk Writing." These scholars describe not only how they teach students writing skills but also how they themselves learned to write and developed their own writing habits. For continuing encouragement and inspiration, as well as observations on the peculiarities of the English language, watch for posts on the *Chronicle's* "Lingua Franca" blog.[2] These successful authors confirm my advice. Create your own rituals. Follow your own rhythm. But do write!

As a corollary, I encourage you to guard your writing time jealously and defend it zealously. During this time, nothing matters more than your book! You can rightfully say "no" to anyone who makes demands on you. If you seek validation for these difficult choices or simply need advice on how to say "no," I recommend *Essentialism: The Disciplined Pursuit of Less* by Greg McKeown, who affirms that you attain your true purpose only by identifying what is essential and deliberately designing your life around it.[3] You chose to become an expert. You chose to write your book. Don't waver!

What can you do when you face a deadline yet you can't summon the urge to write? You might try deception. Tell yourself that you don't have to write that day; you're just going to look at something. Then turn on your computer, open a document, and just diddle around with ideas for 20 minutes. Most likely, diddling will progress to serious writing and, two hours later, you'll discover that you've produced four pages. However, if diddling doesn't fire your writing engine, take a break, then return and try again. After two or three such attempts, if you're still not in the writing mode, perhaps you really should do something else that day. But before you give up in frustration, let me suggest one of my personal morale-boosting techniques, the one that I learned from that undergraduate composition assignment.

Visualize your personal analogy for writing.
Visualize your struggle.
Visualize your triumph.
Then fix your gaze on the image of triumph. And keep writing!

GOOD WRITING

Remember Grandad's home movies? Monticello, 1965. Arcadia National Park, 1967. Chincoteague Island, 1971. The Blue Ridge Parkway, 1974. The real scenery inspires awe, but the images captured by a primitive camera in an amateur's hands that now blur and bump across the screen induce nausea. Sometimes when I read, those home movies come to mind. I glimpse magnificent ideas and provocative arguments, and I lament that the blurry, bumpy writing obscures the author's genius. Your book should convey your expertise on the subject as well as your strengths as a scholar, researcher, analyst, and thinker. But how do you bring your writing skills to the level of your expertise?

Don't bother with style guides; any other genre is a better use of your time. If you want to become a skillful writer, there are only two things you need to do: read and write. Read attentively. Write constantly.

You've received this advice about reading before, along with a mind-numbing catalog of must-reads. In her guide *Reading like a Writer*, for example, novelist Francine Prose instructs you in "close reading"—literally word by word—then hands you a list of over 100 "Books to Be Read Immediately."[4] Resist such assignments! But make your recreational reading as well as your professional reading do double duty. Instead of scanning an article in *International Studies Quarterly*, grazing a *New Yorker* feature story, or racing through a George R.R. Martin fantasy, allow yourself the experience of really reading. Look closely at the author's craftsmanship: his word choices, his verbal imagery, his rhetorical techniques, the structure and length of his sentences, his visual aids, the way he organizes information in paragraphs, building up sections, and then assembling them into a whole.

"Reading" and "writing" cover all media, not just text. The recurring themes in a Haydn string quartet suggest how to keep your argument in the foreground from chapter to chapter. Unexpected word choices in lyrics by David Bowie or Billy Joel make you more daring. After watching a cooking show, you suddenly discover how flavoring each step with a human interest story will hold your readers' attention as you teach them a process. Strolling through an art gallery, you collect a bagful of metaphors to enliven your writing. The best thing about being a writer is that no experience is wasted. Everything—everything—can make you a better writer, if you only pay attention.

Keep a scrapbook of passages that you particularly admire. Savor these pieces. Read them again and again until you understand what makes them work. Annie Proux's clear, simple sentences. Machiavelli's parallel structures. *Scientific American's* graphics. Frederick Forsyth's meticulous description of The Jackal's meticulous preparations for the assassination of Charles de Gaulle. Janet Browne's ability to tuck information about 19th-century science and culture unobtrusively yet precisely where the reader of her Darwin biography needs it. Likewise, make note of writing that you find impenetrable or merely unpleasant. Isaiah Berlin, with his perplexing double negatives and "not that . . . but this" constructions. The overwhelming technical detail in

Andy Weir's novel *The Martian*. The documentary film with too many talking heads.

Don't neglect reading your own work. In particular, reading aloud will help you detect infelicities in your writing.[5] When you have a long break between writing sessions, reading lets you pick up the thread. When you can write continuously, reading the draft introduction from time to time enables you to hold the big picture in mind while you work on the details. Reading your drafts from different angles, even if you are not ready to begin revising them, also allows you to monitor your development as a writer.

You have so much expertise. Have confidence that you'll develop the writing skills to match. Now let's move beyond these generalities to focus on the two elements of good writing: performance and presentation.

PERFORMANCE

Performance, for our purposes, refers to a combination of techniques for holding your readers' attention, communicating with them effectively, and generally providing a pleasant reading experience.

Audience

In one of his routines, recorded on the album *Let's Get Small*, comedian Steve Martin tells a joke involving precise specifications for wrenches and sprockets. Nobody laughs. He thought he was entertaining the National Plumbers' Convention, but he realizes with embarrassment that there are no plumbers in the auditorium.[6] Absurd as it may be, Martin's deliberate mistake highlights the need to adapt your performance to the audience, communicate on their level, anticipate what they don't know, and provide the necessary background. Unless you and your editor agree that you are writing solely for experts in your field, assume that your readers have only minimal knowledge of your subject and that they do not speak your technical language. You may feel as though you are "dumbing down" your work for nonspecialists; but if readers don't understand, they question your intellect—not theirs. Ironically, the better they comprehend your argument and the sophisticated research on which it stands, the more they will recognize your brilliance.

In the 1970s, Steve Martin apparently didn't worry about offending women, Native Americans, fellow performers, senior citizens, or the church-going middle class. Both his material and his delivery, some might say, showed bad taste. While I'm not suggesting that you cater to delicate sensibilities, you want to choose a tone appropriate for your topic and your anticipated audience. For most academic research and scholarship, a straightforward reporter's tone works well. For emotional topics, such as adoption, infertility, or end-of-life care, you might imagine yourself speaking to readers confronting these very issues. For a controversial topic such as immigration, try to control your passion and avoid polemics. When you refute conventional popular or scholarly wisdom, you must refrain from scolding. In every case, you need to treat your readers with respect or they will simply close your book.

The Fatigue Factor

As every teacher and entertainer knows, keeping an audience engaged presents a challenge. The same techniques developed for the stage and the classroom translate to the page. Consider those for preventing fatigue. First and foremost, recognize that time is not on your side. A university class typically runs between 50 and 90 minutes. Have you noticed that concerts, comedy performances, and other forms of live entertainment tend to be divided into similar units? Assume, then, that your readers will allot approximately this amount of time to a single reading session and adjust the length of each chapter accordingly. Chapters containing dense material or passages that require visual imagination, such as battle scenes, force readers to go slowly. Consequently, these chapters should be shorter than those containing generally familiar information, discussion, and anecdotes, which nonspecialist readers can consume quickly.

In Hollywood and Los Angeles, *tempo* refers to the pace at which performers deliver content. In books, it refers to the pace at which readers consume content. Just as you adjust the chapter's length according to the demands it makes on the readers, you want to monitor the sections within the chapter. Consider how much depth you must provide in order to demonstrate your expertise and convince readers of your argument. In any case, avoid page after page of exhausting detail. Interleave description, anecdote, discussion, image, and analysis. Fortify your argument by alternating different types of data in the same chapter. Rather than put all the statistical evidence in one part of your book and all the interview evidence in another part, move back and forth between the two. Intermingle graphics and quotations. Anticipate the appearance of the printed page. Here and there, offer your readers an oasis amidst the monolithic blocks of text. By providing this variety, you slow the inevitable advance of fatigue.

Lively Language

Snappy writing also keeps readers alert. In *The Writer's Diet: A Guide to Fit Prose*, Helen Sword identifies the five key principles of vivacious writing.[7]

* Choose action verbs instead of being verbs (e.g., *am, is, was*).
* Choose concrete nouns instead of abstract nouns.
* Avoid strings of prepositional phrases.
* Use adjectives and adverbs sparingly. By choosing evocative verbs and nouns, you reduce the need for additional descriptive words.
* Beware of *it, this, that, there.* When you must use one of these words, leave no question as to what fact or concept the word refers to.

Speaking from my own life as a reader, an author who applies these principles holds my attention because he engages my imagination. Even nonfiction written in this kind of lively prose draws me into a world I can see, hear, taste, smell, and touch. Surely, you have similar experiences. Collect those passages like souvenirs from your reading travels and revisit them whenever you seek inspiration. Meanwhile, practice these principles in your everyday writing and speaking as well as in your professional writing. Your friends and colleagues will notice your performance!

PRESENTATION

When you designed your table of contents, you spent a good deal of time thinking about presentation. Although you will probably keep the general organization scheme intact as you draft the manuscript, allow yourself the option of rearranging chapters or transplanting sections from one chapter to another. Here I suggest a few strategies for determining how to structure a chapter and where to place particular chunks of information.

Chapter Structure

Remember the formula for the introduction that you submitted with your book proposal? This same formula can serve for individual chapters and the book as a whole. Of course, you are welcome to invent your own formula, but first-time authors seeking a guide, as well as accomplished authors facing a short deadline, rely upon this old standby. Properly used, it operates so quietly that readers don't notice it. Here are the five components of the formula:

* The attention-grabber
* The context and significance
* Your argument
* Your evidence and your study's distinctiveness
* Your findings and their ramifications

Starting on the level of your book, if you look at your table of contents, you will see that your introduction functions as your attention-grabber, and early chapters provide context and reaffirm the significance of your topic. The central chapters contain your argument and evidence, both of which highlight your study's distinctiveness. Your conclusion summarizes your findings and discusses their ramifications.

Turning now to the chapter level, you might open with a startling fact, a provocative statement, a personal anecdote, or an historical moment. Follow your attention-grabber with a few paragraphs outlining that particular chapter and placing it in the context of your main argument. Moving on, develop the chapter's argument, fortifying it with various kinds of evidence. Conclude by restating your findings and connecting them to your main argument in order to set the stage for the next chapter. Please, please, do not copy the chapter introduction and glue it onto the end as your conclusion. Yes, you may repeat key phrases throughout your book but a blatant copy-and-paste from one section of the book to another angers readers—especially if they are also writers—because it looks like cheating or laziness.

Just-in-Time

The five-part formula gives you only a chapter's macro structure; it doesn't tell you how to organize your material within these parts. As I discussed previously, each chapter should contain no more than a nonspecialist can consume in 50 to 90 minutes. At the same time, a chapter should contain only the information the reader requires to comprehend the central idea or argument of that chapter. Apply the Just-in-Time (JIT) approach, first developed by manufacturers who, rather than overstock a warehouse, chose to store only enough inventory to fill customer orders and that now has been adopted in industries as diverse as retail and education. I'll give you an example.

One evening a few years ago, I received a manuscript by e-mail. Scanning the file names of the attachments quickly, I opened the one labeled "Introduction," clicked the print icon, and went to make myself a cup of tea. When I returned a few minutes later, I found an avalanche on my desk and an empty printer bleeping for more paper. That introduction stretched to 105 double-spaced pages. Dreading yet another overgrown manuscript, I opened the remaining attachments and discovered with equal relief and surprise that the page count for the entire manuscript totaled only 322. The next morning, I called the author. After applauding her research and exclaiming over her argument, I asked why she chose to devote nearly one third of her manuscript to the introduction.

"Readers won't know much about India: its colonial history, climate, topography, economy, constitution, religions, and ethnic groups," she answered.

"And unless they understand all these things, they can't understand my argument about the mixed success of development programs. So, I took everything they need to know before they start reading and put it all into the introduction."[8]

"I'm one of those needy readers," I admitted. "Compounding the problem, I have a short memory. There's no way I'm going to recall information that I picked up in the introduction when I need it 50 pages down the road."

Your readers, too, lack essential information. They, too, need it when they need it—not a page earlier or later. Former managing editor of *The Oregonian*, Jack Hart coached many prize-winning journalists, and he packages the same advice he gave them in *Storycraft: The Complete Guide to Writing Narrative Nonfiction*. Although Hart doesn't use the term "Just-in-Time," he devotes most of the chapter on explanatory narratives to the placement of facts, background, explanations, and technical descriptions. Because journalists think of this information as digression from the storyline, Hart recommends sandwiching digressions between narrative slices: narrative, digression, narrative, digression, narrative.[9] If you want models for your scrapbook, you will have to look closely because talented writers slip in information so subtly that casual readers don't notice the detour from the storyline. Historical novelists in particular excel at this sleight of hand, and Hart names his favorite journalists whose work may perhaps interest you.

In your book, which is more complex than a newspaper or magazine feature article, you have more diverse materials to work with than any novelist or journalist: anecdote, argument, quantitative evidence with visuals, qualitative evidence, etc. As you decide how to layer these materials, consider precisely when a reader needs a particular piece of information and how you can employ it to alter the tempo. Returning to the previous example, the snapshot of India's current state of development gave the author an opening for a brief digression on its colonial history. She could've tucked the overview of India's topography into the discussion of programs for building highways and railroads. And her description of India's climate would've fit well into the section on efforts to build the tourist industry because the weather and the threat of tropical disease influences potential tourists' decisions about where and when to travel. Instead of imposing on readers a multi-lecture course, this author could have lured them into her book and educated them gently as they went along if only she had applied the JIT approach.

THE BIG PICTURE

Just as a painter steps back from his canvas from time to time, return to your introduction periodically in order to keep sight of the big picture. Before you start a new chapter, reading the introduction can help you to frame the section on context and significance. After you finish the chapter, a rereading helps you make adjustments so that the chapter carries forward the ideas in the introduction and aligns with your main argument. But be warned: as you think deeply about your research, your argument may evolve. In fact, for a creative, open-minded author, the initial argument often takes an unanticipated turn. Don't be surprised when this happens to you. Savor the thrill of discovery. And resist the urge to rewrite the introduction immediately because your argument will likely leap, twist, or rebound again. Instead, write down the new version of your argument as clearly as possible and post it where you can see it as you work on the manuscript. You want to make a few notes for updating the introduction, but postpone revising it until you've drafted at least half of the manuscript. Alternatively, rather than rework the introduction at that point, you could draft your conclusion, where you fully elaborate your argument and its ramifications. Whatever strategy you choose, by keeping yourself on track as you write your manuscript, you keep your readers on track as they move through your book.

Subheads also help your readers navigate. Rigidly structured outlines, which high school teachers once required for every reading assignment, have gone out of fashion, yet the concept and terminology endure. A formal outline uses capitalized Roman numerals for the top level, capital letters for the next level down, Arabic numbers for the next, lowercase letters for the next, lowercase Roman numerals for the next, and so on. Each successive level is indented one tab farther from the left margin than the level above it. Imagine an outline for your book, with the chapter titles indicated by capitalized Roman numerals. The major sections of a chapter appear at the next level, marked with capital letters (known as *A-heads* in the publishing industry). The divisions of the chapter section are labeled with Arabic numbers (known as *B-heads*). You want to check your publisher's formatting guidelines, but most scholarly presses in the United States ask an author to use only A-heads, typed in bold, flush with the left margin, and B-heads, typed in italics, also flush with the left margin.

You've noticed that in this guide I use mostly A-heads. I reserve B-heads only for long sections that include several subtopics. Taken together, these heads help me not only to follow the discussion but also to spot omissions. Does the chapter on submitting a proposal flow logically? In the chapter on surveying potential publishers, did I mention the benefits and drawbacks of book series? I hope that these subheads allow you to scan quickly for specific information. Could you locate the suggestions for writing a response to reviewers? Could you turn through the publishing contract chapter to find the section on royalty payments? As you decide where to place subheads in your own book, imagine an undergraduate who studied it very thoroughly. An hour before the final exam, she wants to review it. She should be able to flip through the pages, reading the subheads in order to refresh her memory.

Subheads also help you, the author, to track the progression of your argument, spot digressions, and discover disruptions in the logical flow. Before you begin revising your manuscript or even a single chapter of it, take time to read the subheads. Then revise from the macro level to the micro level: the table of contents, the chapters, the subheads in a chapter, the organization of the paragraphs under each subhead. When you're satisfied with the overall structure, you can fine-tune individual sentences.[10]

AGONY AND ECSTASY

What is your personal analogy for writing? Whether you peddle or paddle, break rocks or breakdance, the hard exercise builds your skills and stamina. May the moments of exhilaration fortify you to meet the challenge of persistence! At the same time, I hope that the suggestions offered in this chapter help you overcome the challenges of performance and presentation. Imagine your triumph. Then fix your gaze on that image and keep writing!

NOTES

1. Helen Sword, "'Write Every Day!': A Mantra Dismantled," *International Journal for Academic Development* 21 (August 2016): 312–22; quotation from page 316.
2. Rachel Toor, "Scholars Talk Writing," *The Chronicle of Higher Education*, ongoing series, accessed 29 December 2016, http://www.chronicle.com /specialreport/Scholars-Talk-Writing/26; "Lingua Franca" blog, sponsored by *The Chronicle of Higher Education*, ongoing, accessed 29 December 2016, http://www.chronicle.com/blogs/linguafranca.
3. Greg McKeown, *Essentialism: The Disciplined Pursuit of Less* (New York: The Crown Publishing Group, 2014).
4. Francine Prose, *Reading like a Writer: A Guide for People Who Love Books and Those Who Want to Write Them* (New York: HarperCollins Publishers, 2007).
5. You'll find the advice to read aloud in many places; I'll cite just two: William Germano, *From Dissertation to Book*, 2nd edition (Chicago: University of Chicago Press, 2013), 95; Rachel Toor, interview with Sam Wineberg, "Scholars Talk Writing," *The Chronicle of Higher Education*, 17 August 2015, accessed 29 December 2016, http://www.chronicle.com /article/Scholars-Talk-Writing-Sam/232365?cid=cp26.

6. Steve Martin, *Let's Get Small* (Warner Bros. Records, Inc., 1977).

7. Helen Sword, *The Writer's Diet: A Guide to Fit Prose* (Chicago: University of Chicago Press, 2016); Writer's Diet website, accessed 29 December 2016, http://writersdiet.com/.

8. I've altered the details of this incident in order to protect the author's identity.

9. Jack Hart, *Storycraft: The Complete Guide to Writing Narrative Nonfiction* (Chicago: University of Chicago Press, 2011), 183–202.

10. I learned the macro-to-micro technique from Wayne C. Booth, et al., *The Craft of Research*, 4th ed. (Chicago: University of Chicago Press, 2016), 189–94.

CHAPTER 8 ⸻

Presenting Your Evidence: Quotations, Images, Tables, and Graphs

The author arrived carrying two bulging portfolios containing a collection amassed during five years spent trawling archives and government records for a book on the history of city plans for Washington, DC. Apparently, he envisioned a coffee table book without considering either the high production costs or the comparatively small audience. When I showed him my calculations and explained that the press would require a hefty subvention for such a grand volume, I proposed a compromise: a large but slender book with three dozen black-and-white illustrations and a small gallery of color illustrations. Although he agreed to this scaled-down version, he asked me to help make the selections. Perhaps he hoped that a tour of his collection would induce me to include all of it. Studying the drawings of a city, familiar yet somehow distorted as if it existed in an alternate universe, I began to wonder if I'd underestimated the market demand or the likelihood of obtaining a grant.

But soon I woke to the realization that the majority of the illustrations were iterations of just a handful of master designs, the landmarks in the capital's planning history. Then the author opened the second portfolio. He had letters in 18th-century script, congressional bills, excerpts from 19th-century travelogues, pages from surveyors' notebooks, and portraits of all the principals, from a photograph of U.S. Senator James McMillan to Gilbert Stuart's ubiquitous painting of George Washington. A scholar myself, I understood the author's pride in this collection and his desire to share his fascinating treasures with readers. In my role as an editor, however, I had to point out that a book is not a museum for data, quotations, and images. The exhibits in a natural history museum represent a fraction of its holdings; most lie in storerooms accessible only to researchers. Just as a curator evaluates the collection and judiciously selects items for an informative, theme-based public display, an author must determine which evidence best supports his argument and how to present it most effectively. For this author, the museum analogy struck home because, I learned, he was concurrently assembling an exhibit on the same topic as his book. He lamented that he had been allocated only two small rooms. Whereas designing the exhibit to accommodate the physical space felt limiting, designing it to advance an argument felt liberating. I encouraged him to approach his book in the same way: the argument and the physical space, the ideas and the appearance of the book pages.[1]

You face a similar challenge, with an abundance of evidence and a limited number of book pages. In this chapter, I offer general guidelines for determining which evidence to include and deciding how to showcase it. Because copyright protections may restrict the use of certain materials, I begin with an overview of the fair use provisions of U.S. copyright law. Next, I discuss considerations for selecting the most compelling quotations and images and incorporating them into your text. I then show you how to display your numeric data in readily comprehensible tables and graphs. By applying these guidelines, you will deploy your evidence to advance your argument most effectively.

COPYRIGHT AND FAIR USE

Whatever kind of material you want to reproduce, whether it is a quotation from a novel, a line from a musical score, or a section of a painting, you need to ascertain if the item (aka, the work) is under copyright and what restrictions may apply to its use. In the course of working with hundreds of authors just like you, I developed the following evaluation process using a series of questions somewhat like a flowchart or decision tree.

Is the work in the public domain?

Works that were never under copyright, such as the U.S. Constitution and many other government documents, as well as works whose copyright protections have expired, such as the first edition of James Fenimore Cooper's *Last of the Mohicans*, are in the public domain. No copyright restrictions apply, and you do not need to obtain permission to reproduce any part or any proportion of them. If you gathered material from a curated collection, however, you might ask the archive, museum, or library for permission as a courtesy.

Did the creator apply a Creative Commons (CC) license to the work?

A CC license complements copyright. Whereas copyright explicitly restricts rights, a CC license allows the creator to make his work freely available and specify precisely how it may be used. The creator may require attribution, meaning that the user must acknowledge the creator. He may forbid the commercial use of the work or permit both commercial and noncommercial uses. He may prohibit derivatives, insisting that the user maintain the integrity of the original work; or he may allow derivatives so that the user can alter and remix the work to produce something new. Finally, employing the "share alike" provision of the CC license, the creator can demand that the user make the resulting new work available under a CC license with the same terms.[2]

If you wish to reproduce a work or a portion of a work bearing a CC license, you do not need the creator's permission, but you should adhere to the license's terms. Typically, reproduction in a scholarly or professional book is considered a noncommercial use, unless you plan to use the work on the book's cover or for marketing purposes. Also, be aware that your publisher cannot comply with a "share alike" mandate.

Is the work under copyright? Do the fair use provisions apply?

First, recall that a work is under copyright even if the creator did not officially register the work with the Copyright Office. When you reproduce a portion of a work that is currently under copyright, you may be eligible for the fair use exemption. This provision of the *U.S. Copyright Act* allows reproduction for the purpose of "criticism, comment, news reporting, teaching (including multiple copies for classroom use), scholarship, or research."[3] To determine whether or not this provision applies to a particular case, the law employs four criteria:

(1) the purpose and character of the use, including whether such use is of a commercial nature or is for nonprofit educational purposes;
(2) the nature of the copyrighted work;
(3) the amount and substantiality of the portion used in relation to the copyrighted work as a whole; and
(4) the effect of the use upon the potential market for or value of the copyrighted work.[4]

In short, the fair use provision allows for reproduction of a small portion of a work under copyright for the purpose of scholarly research and critical commentary, as long as this reproduction does not reduce the commercial value of the original work. Quoting a significant percentage of a poem could reduce the commercial value of the volume from which it came because a reader might conceivably choose to purchase your book rather than the poet's. Reproduce only as much of the work as necessary in order to make your point.

If you use copyrighted sources extensively, I suggest that you spend some time with one of the excellent introductory guides available in print or online. The College Art Association, the American Musicological Society, and many other professional societies also publish guidelines regarding fair use. When in doubt, don't hesitate to consult your acquisitions editor, who deals with questions of copyright and fair use daily.[5] Should you determine that the fair use provision covers your use of a particular work, you do not need to seek permission from the copyright holder.

Fair use does not apply. Will the copyright holder grant permission for the intended use of the work?

If you determine that the fair use provision does not apply, you need to seek permission from the copyright holder. When you approach him or her (or, in some cases, an organization), identify the work which you want to use (e.g., a painting) and how much of the work you wish to reproduce (e.g., a paragraph from a short story), for what purpose (e.g., a scholarly article or book), and the medium or format (e.g., print-only book, online/digital publication) in which your own work will appear. For a snippet of a musical score or a line from song lyrics, clarify that you request permission to reprint this excerpt in a scholarly book—not to perform or record the piece. The guides in the previous endnote include sample permission request forms, although I urge you to use your publisher's standard form to ensure that you obtain exactly what's required. Some copyright holders may insist that you use their permission request form, but before you submit such a document, consult your acquisitions editor. You don't want permission restrictions to limit the publishing options for your book!

Regardless of the laws or licenses governing an item's use, you need to cite it properly. In all fields, professional ethics demand honesty in reporting the results of one's work and in claiming credit for it. Plagiarism carries the same taboo as the fabrication or falsification of data. The U.S. Federal Research Misconduct Policy, which applies to all federally funded projects, defines research misconduct as "fabrication, falsification, or plagiarism in proposing, performing, or reviewing research, or in reporting research results." The policy then defines plagiarism as "the appropriation of another person's ideas, processes, results, or words without giving appropriate credit."[6] In short, scholars and researchers consider plagiarism a serious offense. Cite your sources.

Although citing too much does not carry the same penalties as plagiarism, it can lead readers to question an author's maturity, if not his integrity. Overly zealous students enumerate every source encountered in the course of their research and further pad their theses with sources unearthed in a database search but never actually read. Too often, I've seen these student habits persist among mature scholars and researchers. You're an expert. You do thorough research, present your evidence honestly, and properly acknowledge the ideas and contributions of others. Your peers trust you. Resist the compulsion to cite everything. Instead, limit your citations to the sources that you actually engage with.

QUOTATIONS

A self-righteous but nonetheless terrified adolescent, I filled high school term papers with quotations to show off my research. This tactic earned me a reputation as a studious student, so I used it again in college, with equal success. But when I arrived in graduate school, I began to notice how the scholars whom I most admired strategically incorporated quotations into their writing. In fact, they seldom used quotations. Instead, they summarized or paraphrased the scholarly literature as well as the sources that they had collected themselves. They extracted expressive quotations from interviews, letters, speeches, reports, and other unique documents, which they artfully wove into their own sentences. Only occasionally, when they wanted to reveal a person's character or capture an eyewitness scene, did they insert a block quotation. Framed with white space in the center of a book page, such a quotation really did resemble a portrait or a photograph. By imitation, I learned how to use quotations effectively. And I still believe that imitating masters of the craft is the best way to learn every writing skill. If you want an authoritative treatment of the rules for summarizing, paraphrasing, and quoting, you can pick up one of the many guides for academic writing.[7] But with your indulgence, I will provide a quick, irreverent review.

When you are simply harvesting information, a summary works best. Rather than repeat the entire folksong "Old Lady Who Swallowed a Fly," for example, you could summarize it thus: an elderly woman died after gulping a succession of animals.[8] When you want to stay close to the original but its wording seems arcane, laborious, or otherwise inappropriate for your audience, paraphrase it. Paraphrasing the nursery rhyme "Little Miss Muffet," you get: Sitting on her stool, Little Miss Muffet was eating curds and whey when a spider landed next to her and gave her a scare.[9] When you cannot alter the wording of the original without changing its meaning, as in the case of legal statements, or when the wording itself makes your point in a startling way, place it in a run-in quotation. I'd argue, for example, that silly songs deliberately corrupt a child's education. How many youngsters misspell *farm* after singing "Old MacDonald had a farm, E-I-E-I-O"?[10]

When you can't improve on the original wording to sketch a character or a scene, use a block quotation. Think of it as an image, with white space as the frame and your introduction as the caption, explaining what the quotation is about and why it matters. At the end of the quotation, add the source citation.

The following excerpt from Lewis Carroll's *Through the Looking Glass, and What Alice Found There* merits a block quotation both because changing the words even slightly would alter the meaning and because the excerpt vividly captures a scene:

> 'T was brillig, and the slithy toves
>   Did gyre and gimble in the wabe;
> All mimsy were the borogoves,
>   And the mome raths outgrabe.[11]

Clearly, the opening stanza of the irreducible, inimitable, and unforgettable "Jabberwocky" defies summary or paraphrase.

You do not need permission to summarize or paraphrase a work, although you may need permission to quote from it. If it is in the public domain or if the creator has applied a Creative Commons license, it isn't necessary to obtain permission for a quotation. If the work is under copyright, determine whether or not the fair use provision applies. For quotations drawn from letters, diaries, records, or other unpublished materials, check the policy of the archive in which you found them. It may be possible to secure a blanket permission for all the collections that you cite extensively. For quotations from interviews conducted by others, the recommendations noted earlier apply, whether these interviews have been published or preserved in an archive. For quotations from interviews that you conducted yourself, you have the release forms signed at the time of the interview.

IMAGES

In your research, you gathered a number of intriguing images; but you have space in your book for only 10 percent of them. How do you decide which to include? Following a process of elimination makes the decision less painful. First, weed out photographs of newspaper clippings, letters, receipts, and diary entries. A summary or a quotation better serves your purpose than an illegible reproduction of a document. Comic strips and political cartoons likewise lose their luster on a book page, and obtaining permission to reprint them poses a challenge. Treat the comic strip as an anecdote, or describe the cartoon and engage your reader's imagination. If the image is available on the web, provide a link in the citation.

Second, eliminate images that don't meet your publisher's quality specifications regarding dimensions and resolution in dots-per-inch (dpi). A photograph that appears cramped, bloated, fuzzy, or pixelated on your screen will look worse on a book page. Keep in mind that unless the print version will include color, the designer will convert your drawings and photographs to halftones. It's difficult to argue for the significance of the Horse of a Different Color in the Wonderful City of Oz when your supporting evidence appears in black and white.

Indeed, every one of your images serves as evidence for your argument. An image should be illustrative, not merely decorative. Set aside all images that fail this criterion. Now look closely at those that remain. Could you enhance any of them so as to make your point more readily apparent? The man poised to whip his horse hides in the background in Sandro Botticelli's crowded *The Adoration of the Magi*, and he will practically disappear when the painting is reduced to fit a book page. You can make sure that the reader sees him by lifting out a detail centered on the horseman or cropping the painting vertically along the ruin's right wall. In scientific illustrations, authors regularly insert a measure of scale next to foraminifera, aim an arrow at Ganymede, or circle the abnormal leukocyte. Could you do something similar?

Your caption can further connect the image to your argument. Here, for example, is my caption for *Adoration:* Whip in hand, a man stands poised to strike the rearing horse. This figure foreshadows the Crucifixion of Christ who challenged Roman and Jewish authority. Detail from Sandro Botticelli, *The Adoration of the Magi*, c. 1478/1482, Painting, Courtesy National Gallery of Art, Washington, accessed 4 December 2016, http://www.nga.gov/content /ngaweb/Collection/art-object-page.24.html. This caption addresses the three essential questions: 1) What is the reader is looking at? 2) Why does it matter? 3) Where does it come from? By providing a description of the image, you make your book fully accessible to visually impaired readers who use text-to-voice software which cannot interpret images. By explaining why the image matters, you point out its significance as evidence for your argument. Finally, by citing the source, you validate the evidence and acknowledge the individual or institution permitting you to include the image.

Copyright law regarding works of art is too complicated for a responsible summary here. I recommend the College Art Association's *Code of Best Practices in Fair Use for the Visual Arts* and, as always, consultation with your acquisitions editor.[12] In my experience, archives respond generously to requests, whereas media conglomerates often deny permission or charge extraordinary prices. Professional artists and photographers typically fall somewhere between these two extremes. When you approach the copyright holder, attach a copy of the image, state where you found it, and provide information about its origin (e.g., title, creator, date). Indicate where you intend to reproduce the image (e.g., a scholarly article or book) and the medium or format (e.g., print-only book, online/digital publication). Also, inquire if there are any restrictions on cropping, inserting pointers, or otherwise altering the image. As I explain earlier, you would be wise to use your publisher's permission request form to ensure that you secure the necessary rights. If a copyright holder insists that you submit a different form, consult your acquisitions editor lest idiosyncratic restrictions force your publisher to rescind an image from digital formats or future editions of your book.

Thankfully, the Metropolitan Museum of Art, the Getty Museum, the National Gallery of Art, and several other leading museums around the world now provide photographs of works in their collections for download and publication at no charge.[13] The U.S. Library of Congress and the various museums comprising the Smithsonian Institution feature an abundance of public domain images. Also, Creative Commons hosts a directory of images and other materials which their creators offer under open access licenses.[14]

And now for the fun part—deciding how to arrange the images in your book. Obviously, you want to place them beside the relevant text. For a single image, you simply indicate where it belongs in your manuscript. Your publisher's guidelines should explain how to insert these callouts. But if the synergy of juxtaposed images exceeds the impact each has on its own, you might consider arranging them in a gallery. Resembling an art exhibit, a gallery consists of several pages devoted exclusively to images. As the curator, you create an exhibit that conveys your argument, or at least one aspect of it. You often find such a visual essay in biographies and studies focused on change over time as well as architectural surveys and similar comparative studies of places, objects, and organisms. In addition to these rhetorical benefits, a gallery offers an aesthetic benefit because images scattered throughout your book will appear on standard paper but your publisher may choose high-quality photo stock for the pages in your gallery.

MAPS

A map enables your readers to visualize the space you write about and the relationships among features within it. This is why fantasy author Brian Ruckley draws maps of Haig lands for his epic series and journalist Ted Conover sketches a map of Sing Sing for his report on the time he spent working there undercover as a correction officer.[15] Appearing in the first few pages, such a map serves as an easy-to-find reference and makes a handsome frontispiece. On the other hand, if you have a number of maps, consider distributing them throughout the book so that each appears exactly where your reader needs it most.

The criteria for selecting images apply equally to maps. Choose only those which most effectively illustrate your argument. You want to avoid littering your book with common maps, such as a state map of the United States, unless you emphasize particular features that are not commonly depicted, such as nuclear power plants and wind farms. In the latter case, you may be able to modify an existing map; but please make sure it meets your publisher's specifications. Alternatively, you might hire a professional mapmaker or recruit an experienced colleague.

In addition to a legend identifying the symbols, a map requires a title telling the reader what she's looking at and a caption explaining its significance and citing its source. For permissions, treat a map as you would any image. The U.S. Library of Congress, Yale University Library, and the U.S. Geological Survey are only a few of the libraries and government agencies that offer a fabulous collection of maps in the public domain.[16]

VISUAL PRESENTATIONS OF NUMERIC DATA

Sometimes I need to hear my own advice. For years, as an acquiring editor in the social sciences, I warned authors about overstuffing their books with tables and graphics. I couldn't understand why ranks of numbers and stark diagrams intoxicated such intelligent, and otherwise sober, researchers. Perhaps *intoxicated* overstates the case but at least their judgment seemed impaired.

Then, suddenly given responsibility for a new digital repository, I had to prepare semiannual reports documenting progress, as demonstrated by the collection's growth and web visitors' use of it. I didn't have much data for my first report, but I parsed it every which way and generated what I considered an impressive set of graphics. The administration's response amounted to "Well, we've made a good start." Seeking more critical feedback, I consulted a colleague whose clear, concise analytical reports I admire. He spent more than an hour reviewing my pathetic little report and patiently explaining how to capture the right data and design a visual presentation that conveys the right point. I left his office full of gratitude—and extremely embarrassed. Essentially, he had reiterated the exact same advice I've been giving authors. Yet now I understand how knowing that the merit of your work is judged on the basis of your data, combined with the alluring capabilities of software, leads you to fill pages with fancy, important-looking tables and graphics. No matter if they are redundant and relatively meaningless. As long as they mesmerize the reader, they serve the purpose. But you don't need mesmerism. You are an expert. You've conducted extensive research, thoroughly analyzed your evidence, and built a persuasive argument. You would do better with tables and graphics that your readers will instantly comprehend and always remember. So, with humility and empathy, I offer the following recommendations for creating such visuals.

Before turning to the specifics of tables and graphics, let's consider the basic principles for choosing which of these two is most appropriate in a given situation. First, when you have a few simple numbers, summarize them in prose. When you have many complex numbers and their precise values matter, display them in a table. When the relationships among the numbers or the trends across the numbers are what matter, draw a graphic. As a general rule, keep your tables and graphics as simple as possible; design them so as to make your point immediately evident. In other words, think of tables and graphics as visual presentations of your argument.[17] Familiarize yourself with conventions in your field, but also spend some time with publications in neighboring fields—and even distant fields—for inspiration. Create a scrapbook with examples you admire. Who knows? A chart in an astrophysics journal may suggest an innovative way to represent interactions among youth in fanfiction clubs. Although you can probably use common software such as Microsoft Excel, check your publisher's guidelines to make sure you provide tables and graphics in the required formats.

TABLES

The orderly, no-nonsense appearance of a well-designed table disguises the creativity behind it. In many ways, the process of drafting a table resembles solving a mechanical puzzle like Rubik's Cube. I can't give you step-by-step instructions because only you know what puzzle you have to solve (i.e., the point you want to make) and only you can recognize the solution (i.e., a table that efficiently conveys that point). Nonetheless, I refer you to *The Chicago Guide to Writing about Numbers,* in which Jane E. Miller helpfully explains the logic for framing a table and organizing the interior compartments. She also describes the various types of tables and their uses.[18] I strongly recommend her chapter on tables to authors who write about numbers, whether they run statistical analyses of capital bond markets or count 19th-century tombstones in a New England parish. Here I summarize Miller's lesson.

Each table should address a single point as thoroughly as possible. As you decide whether to consolidate or split tables, you want to take into account the size of the typical book page. How will this table look when the designer shrinks it to a 6"×9" sheet of paper? E-books do not eliminate this problem. In fact, they exacerbate it because each table has to be resized for the particular device that the reader is using.

Miller uses the phrase "anatomy of a table" to invoke the image of a body. Emphasizing that form must match function, she examines the parts of a table, one by one.[19]

Title

State the obvious; it may not be so obvious to your reader. Tell him what he's looking at and what he's supposed to see.

Rows and Columns

Think carefully about which data belong in the horizontal rows and which in the vertical columns. As a general rule, the critical comparisons belong in columns, whereas similar elements belong in rows. Experiment with the order of the elements in the rows and the columns until you find the arrangement that most strongly conveys your point. Don't simply default to alphabetical order. For a complex table, you might consider grouping columns and rows with spanners. Label the columns and the rows clearly, and don't forget to add units of measure. You want to use a similar scheme for all your tables so that the reader can easily interpret them individually and spot trends across them.

Interior Cells

The numbers within the cells should only be as precise as you need them to be.

Captions and Footnotes

The caption for each table should name the source of the data. If you gathered the data, you do not need permission either for the data or for the table. If you use data gathered by another researcher, you need to cite the source and specify that you're using only the data. (Data are facts, and facts are not eligible for copyright protections.) If, however, you reproduce a table from another source, you must obtain permission from the copyright holder and provide a citation, including the phrase "used with permission." Definitions, lists of abbreviations, and comments on the data belong in the table's footnotes.

The best table speaks for itself. Although you discuss your tables and the meaning of the data in the text, each table should provide enough information that the reader can interpret it—and grasp your point—without reference to the text.

Table 8.1. This exemplary table efficiently presents information and effectively reinforces the author's argument.

Table 2.1   Firearms Imported into the United States, by Country of Manufacture, 2013

	Handguns	Rifles	Shotguns	Total Firearms
Brazil	452,165	404,234	119,090	975,489
Austria	932,117	21,653	618	954,388
Germany	518,150	135,381	1,370	654,901
Italy	237,918	53,115	212,557	503,590
Croatia	451,657	0	0	451,657
Turkey	105,757	0	306,312	412,069
Canada	6,030	369,512	5	375,547
China	0	4,155	234,486	238,461
Russia	772	169,112	34,904	204,788
Philippines	140,813	5,909	9,800	156,522
Serbia	50,658	44,672	0	95,330
Argentina	82,635	0	0	82,635
Japan	0	76,399	1,525	77,924
Czech Republic	39,897	26,856	142	66,895
Romania	3,655	44,734	0	48,389
Belgium	14,499	29,920	10	44,429
Finland	0	43,858	0	43,858
Israel	23,979	18,504	0	42,483
Bulgaria	8,397	31,087	0	39,484
Spain	262	17,760	1,620	19,642
United Kingdom	92	4,345	7,204	11,641
Switzerland	5,806	3,890	0	9,696
Poland	8,406	510	0	8,916
Portugal	20	4	6,415	6,439
Ukraine	4,000	0	0	4,000
Korea	3,879	0	0	3,879
Other	1,401	770	177	2,348
Taiwan	0	1,396	0	1,396
United Arab Emirates	1,359	0	0	1,359
Slovak Republic	1,204	0	0	1,204
Totals	**3,095,528**	**1,507,776**	**936,235**	**5,539,539**

Source: Department of Justice, *Firearms Commerce in the United States Annual Statistical Update 2014*.

Source: Reproduced from Sirgo, "U.S. Import and Export of Firearms" (2015). Used with permission.

This table, reproduced from Henry B. Sirgo's chapter in the three-volume *Guns and Contemporary Society: The Past, Present, and Future of Firearms and Firearm Policy*, exemplifies the proper application of these guidelines. In Sirgo's text, this is Table 2.1 because it is the first table in the second chapter of the volume in which it appears. The title "Firearms Imported into the United States, by Country of Manufacture, 2013" leaves no question as to the content and scope of the data; and the caption specifies the source, "Department of Justice, *Firearms Commerce in the United States Annual Statistical Update 2014*." The reader's eye moves from the top to the bottom of the page, so the table lists the countries of origin in the first column, in descending order, from Brazil through the Slovak Republic. The other columns list three types of imported firearms and the total for these three columns. Here again, because the reader's eye moves from left to right, the category with the highest numbers—handguns—appears first, followed by rifles and shotguns. The interior cells contain whole numbers; note that there is a "0" in a cell when a country did not sell a particular type of firearm to U.S. buyers. All the numbers tally neatly in the cell at the bottom-right corner. Even at a glance, the reader can grasp the astonishing number of imported weapons. This evidence adds detail and impact to Sirgo's observation: "Imports of firearms into the United States generally go to the civilian population, which is far more heavily armed than the population of any other affluent nation."[20]

GRAPHICS

When you want to illustrate a pattern or highlight a trend, draw a graphic. But should you choose a bar graph? A pie chart? A line graph? A scatterplot? In *The Craft of Research*, Wayne C. Booth and his co-authors describe the different types of graphics, the nature of the data they can present best, and their rhetorical uses. I especially recommend the authors' handy summary table.[21] For a more advanced lesson, I refer you to Miller's *The Chicago Guide to Writing about Numbers*.[22] As before, I don't attempt to replicate her eloquent instructions, but rather draw upon them as I outline the fundamentals. Let's first look at the factors to consider when determining which type of graph to use in a particular instance.

Key Point

Choose the graphic that best presents your evidence in a persuasive, easily comprehended way and, thus, supports your argument.

Number and Type of Variables

How many dimensions of your data do you want to depict? Do they concern individual quantities, proportions of a whole, distribution of a quantity, or change over time or space? A line graph, for example, traces a trend for a single variable. A pie chart displays only one variable and its distribution. A bar graph invites comparisons of quantities across time, space, or category. A stacked bar graph combines the advantages of the latter two, distribution and comparisons. That is, imagine creating a pie chart for each of your samples then squashing each pie into a column and arraying the resulting columns along the X-axis (i.e., horizontal axis) of your graph. Adjust the scale of the Y-axis (i.e., vertical axis) so that the reader can easily see the units of measure produced by each slice of each column.

Scale and Scope of the Data

When a graphic cannot accommodate your data, you have several options. You could put your data into a table or multiple tables. You could reduce the scope (i.e., the range) of the data or change the scale (e.g., from meters to kilometers, from yearly intervals to five-year intervals) in a single graphic. Alternatively, you could split that single graphic into a series. Be aware, however, that seemingly redundant graphics clutter your book. Choose the option that makes your point most succinctly.

Audience

What kinds of data visualization are your readers most familiar with? Whereas the math-based sciences and social sciences continually add to their expansive and highly sophisticated repertoire, the humanities generally rely on basic line graphs, bar graphs, and pie charts. You want to communicate in a visual language that your readers understand.

Appearance of the Printed Page

Keep in mind that your designer will resize the graphic to fit one book page, perhaps even half a page. Whereas many e-book formats display graphics in color, you should anticipate that the print book will contain only black-and-white images. Will your graphics, with all their numbers and symbols, be legible when they are reduced and converted to grayscale? Scatterplots, for example, are particularly problematic because the little dots merge into a dark mass.

Limitations of the Software

Your publisher may prefer simple graphics software, such as Excel, to ensure compatibility with all the systems involved in publishing a book. Can you use this software to generate the graphics you envision?

Regardless of the type, a graphic must stand on its own. Although you introduce it and explain what it means in prose, your reader should be able to interpret it—and immediately grasp your point—without referring to the text. For this reason, each graphic needs a title, labels clarifying what is being measured and the unit of measure, and a caption stating the source of the data. A complex graph also needs a legend identifying the symbols.

As with tables, if you draw a graphic using your own data, you don't need permission either for the data or for the graphic. If you use another researcher's data, cite the source. If you reproduce another researcher's graphic, however, you must obtain permission from the copyright holder and provide a citation, including the phrase "used with permission."

This line graph, reproduced from *Climate Change: Examining the Facts* by Daniel Bedford and John Cook, illustrates the ways in which an effective graphic can simultaneously condense your data and underscore your point. Labeled Figure 2.3 in the original publication, the graph compares the global average temperature (scale in degrees Celsius on the left Y-axis) with the annual average number of sunspots (scale in integers on the right Y-axis) from 1850 through 2010 (scale in years on the X-axis). The authors not only provide a legend for the four symbols used on the graph, they also provide an explanatory caption and specify the sources of their data. I have not reproduced the other graph (Figure 2.2) to which the authors refer, but I should mention that TSI is an abbreviation for total solar irradiance. In the text, Bedford and Cook write "sunspots and global average temperatures track quite closely up until the late 1950s. At this point, the Sun started to become slightly less bright—but global average temperatures continued to climb." On the graph, the reader immediately notices the increasing divergence between the temperature trend and the sunspot trend over time. In this way, the authors have made the evidence for their claim both accessible and convincing.[23]

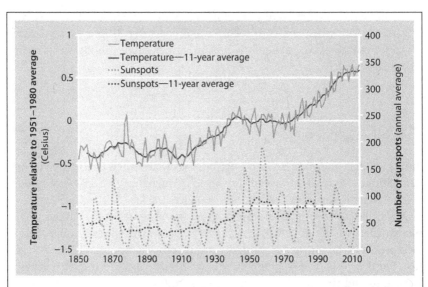

Figure 2.3 Annual average number of sunspots and global average temperature. Note that sunspots and temperature are closely related until the late 1950s, when sunspots begin to fall but temperature continues to rise. Sunspots are a good indicator of TSI, as shown in Figure 2.2. Temperature is shown as difference relative to 1951–1980 average, in Celsius. Eleven-year averages are shown to smooth out short-term variations, bringing the patterns into focus.

Source: Temperature data from Berkeley Earth (2015); sunspot data from WDC-SILSO (2015).

Figure 8.1. This exemplary graph efficiently presents information and effectively reinforces the authors' argument. Reproduced from Bedford and Cook, *Climate Change* (2016). Used with permission.

SHARING THE THRILL

In this chapter, I provided general guidelines for using quotations, images, maps, tables, and graphics to present your evidence. Depending on your field and topic, you may want to incorporate additional kinds of material that I did not cover. For static evidence, such as a diagram depicting a process, apply the previous guidelines for images. For dynamic or interactive evidence, such as video, audio recordings, or games, talk with your acquisitions editor about linking to external websites or posting files on the press website. Be creative!

Although I've repeatedly emphasized the importance of advancing your argument, truly compelling evidence does something more. It invites your readers to share your delight in research and discovery. Recall your sense of vindication when, after four days in the archive, you found a collection of letters from Pancho Villa's henchman to his mother. Properly framed, a few quotations from those letters give readers the same front-row seat to both the Mexican Revolution and your historical investigation. Remember that moment of revelation when you discerned patterns in your data? Tables and graphs can make those unsuspected patterns instantly apparent—but no less astonishing—to readers. And don't forget when you suddenly noticed the horseman, whip in hand, in Botticelli's *Adoration*. Point him out to your readers, and let them see for themselves how cleverly the Renaissance painter captured the trajectory of the Christ Child's life. Experience is the most convincing kind of evidence.

NOTES

1. I changed the details of this incident in order to protect the author's identity.
2. Creative Commons, accessed 8 March 2017, https://creativecommons.org/.
3. *U.S. Copyright Act*, 17 *U.S.C.* § 107, accessed 30 November 2016, http://www.copyright.gov/title17/.
4. Ibid.
5. Kenneth D. Crews, *Copyright Law for Librarians and Educators: Creative Strategies and Practical Solutions*, 3rd ed. (Chicago: American Library Association, 2012); Kenneth D. Crews and Dwayne K. Buttler, "Fair Use Checklist," accessed 30 November 2016, https://copyright.columbia.edu/basics/fair-use/fair-use-checklist.html; Copyright Advisory Office, Columbia University Libraries, accessed 30 November 2016, https://copyright.columbia.edu/about.html; Copyright & Fair Use, Stanford University Libraries, accessed 30 November 2016, http://fairuse.stanford.edu/; College Art Association, *Code of Best Practices in Fair Use for the Visual Arts* (2015), accessed 3 December 2016, http://www.collegeart.org/fair-use/; American Musicological Society, "Best Practices in the Fair Use of Copyrighted Materials and Music Scholarship" (2010), accessed 8 March 2017, http://www.ams-net.org/AMS_Fair_Use_Statement.pdf.
6. Federal Register: December 6, 2000 (Volume 65, Number 235), 76260–76264, accessed 13 September 2016, http://ori.hhs.gov/federal-research-misconduct-policy.
7. Wayne C. Booth, et al., *The Craft of Research*, 4th ed. (Chicago: University of Chicago Press, 2016); Kate L. Turabian, *A Manual for Writers of Research Papers, Theses, and Dissertations, Chicago Style for Students & Researchers*, 8th ed., revised by Wayne C. Booth, Gregory G. Colomb, Joseph M. Williams, and the University of Chicago Press Editorial Staff (Chicago:

University of Chicago Press, 2013); Miguel Roig, *Avoiding Plagiarism, Self-Plagiarism, and Other Questionable Writing Practices: A Guide to Ethical Writing* (2003, revised in 2006 and 2015), accessed 29 November 2016, http://ori.hhs.gov/avoiding-plagiarism-self-plagiarism-and-other-question able-writing-practices-guide-ethical-writing.

8. "An Old Lady Who Swallowed a Fly" (The British Council, 2012), accessed 29 November 2016, www.britishcouncil.org/learnenglishkids.

9. "Little Miss Muffet," accessed 29 November 2016, http://www.rhymes.org .uk/little_miss_muffet.htm.

10. "Old McDonald Had a Farm" (The British Council, 2012), accessed 29 November 2016, www.britishcouncil.org/learnenglishkids.

11. Lewis Carroll, *Through the Looking Glass, and What Alice Found There,* illustrated by John Tenniel (n.p.: Thomas Y. Crowell & Co., 1893), 31–33; accessed 29 November, www.hathitrust.org.

12. College Art Association, *Code of Best Practices.*

13. Metropolitan Museum of Art, http://www.metmuseum.org/about-the -met/policies-and-documents/image-resources; Getty Museum, http:// www.getty.edu/about/opencontent.html; National Gallery of Art, http:// americanhistory.si.edu/collections, all accessed 3 December 2016.

14. U.S. Library of Congress, https://www.loc.gov/collections/; National Museum of Natural History, Smithsonian Institution, http://naturalhistory .si.edu/rc/; National Museum of American History, Smithsonian Institu- tion, http://americanhistory.si.edu/collections; Creative Commons, https:// search.creativecommons.org/, all accessed 3 December 2016.

15. Brian Ruckley, *Winterbirth, The Godless World Trilogy: Book One* (New York: Orbit, 2006); Ted Conover, *Newjack: Guarding Sing Sing* (New York: Knopf Doubleday, 2001).

16. U.S. Library of Congress, https://www.loc.gov/maps/collections/; Yale University Library Map Collection, http://web.library.yale.edu/maps; U.S. Geological Survey, https://www.usgs.gov/; see the list, which is by no means comprehensive, posted on Public Domain Sherpa, http://www .publicdomainsherpa.com/public-domain-maps-resources.html, all accessed 8 March 2017.

17. Booth et al, *Craft of Research*, 214–31.

18. Jane E. Miller, *The Chicago Guide to Writing about Numbers: The Effective Presentation of Quantitative Information*, 2nd ed. (Chicago: University of Chicago Press, 2015), 121–49.

19. Ibid.

20. Henry B. Sirgo, "U.S. Import and Export of Firearms," in *Guns and Con- temporary Society: The Past, Present, and Future of Firearms and Firearm Policy,* ed. Glenn H. Utter, Vol. 2 (Santa Barbara, CA: ABC-CLIO, LLC, 2015): 51–74, quotation, 57; table, 60.

21. Booth et al., *Craft of Research*, 230–31.

22. Miller, *Writing about Numbers*, 150–90.

23. Daniel Bedford and John Cook, *Climate Change: Examining the Facts* (Santa Barbara, CA: ABC-CLIO, LLC, 2016): 67–68.

CHAPTER 9 ───────────────────

Preparing Your Final Manuscript: The Components of a Book

Just as you deployed a charm offensive targeting acquisitions editors, reviewers, and editorial committees to win a publishing contract, now as you prepare your final manuscript, you want to launch a second offensive, this time targeting your publisher's production team. Indeed, your relationship with the team affects not only your experience as an author but also the success of your book.

But how do you charm copyeditors, designers, and production managers? First and foremost, you need to realize that they view themselves as artisans. Despite the new digital technologies that have transformed every aspect of the industry over the past three decades, these artisans continue to treat the publication process as a ritual and the book itself as a ritual object. Publication still requires a series of movements executed in a precise sequence by skilled practitioners to produce an object with multiple parts in a traditional arrangement. In a word, publishing professionals value order. By following your publisher's submission guidelines and the instructions given here, you show respect for your team members and demonstrate that you, too, value order. Believe me, with a neat, complete manuscript, you will score points.

In this chapter, I walk through all the parts of a book from front to back. (Your book may not include every one of these parts.) Skimming this chapter, you can use the subheads as a checklist when you prepare your final submission. Reading this chapter a bit more closely, however, you'll come to view a book's mechanics as publishing professionals do. By presenting yourself as a knowledgeable colleague, you can prevent delays, misunderstandings, wasted effort, and innumerable frustrations.

FRONT COVER

Although the publisher has the last word on the cover design, your acquisitions editor will most likely solicit your ideas. If your book will appear in a series with a standard design, you'll not have much input. If it will appear as an independent title, however, you may have more involvement. For inspiration, survey recently published books in your field and neighboring fields for common themes, (e.g., typographical designs, favored colors, abstract images). On the one hand, you want to avoid clichés such as a man-behind-the-podium photo representing election campaigns or images of medical equipment symbolizing healthcare. On the other hand, you want a design that instantly conveys your book's general subject. A bicycle, for example, doesn't suggest interplanetary exploration.

Image

First, discuss with your editor which is preferable: a typographical or an image-centered design. In my experience, a typographical design works best for a theoretical book because conceptual arguments are not easily represented by material objects. But a simple image may also serve. Visit a stock photo site and run a keyword search. These sites also host an abundance of images that invite creative interpretation, so try searching with terms such as *lines, patterns, marbles abstract,* or *architecture abstract.* Once you figure out how the site tags images, you'll discover plenty of options.

If you want a specific place, person, or object, a stock photo website may have it; but you're more likely to find what you're looking for in an archive, museum, or library. Be aware that many of these institutions charge not only for the image file but also for permission to use it. Furthermore, do not be surprised when they charge a higher fee for an image used on the cover than for one placed inside the book. The cover image supposedly has a higher commercial value because it helps to sell the book. Fortunately, the U.S. Library of Congress, presidential libraries, and many other government institutions, as well as a growing number of cultural institutions, including the Metropolitan Museum of Art, have created open access collections of high-quality images that can be downloaded and used without restriction.[1] Unless you are a professional artist or photographer, I don't recommend submitting your own work. The piece that looks so handsome on your wall or your screen could appear quite amateurish on your book.

As you evaluate images, consider orientation, scale, and density of detail. A good candidate has a vertical orientation. Choose the giraffe, not the crocodile. A good candidate also scales well to the dimensions of a book. Choose a Paris cathedral or just a few spires, not the panoramic cityscape. Small features will disappear when a designer reduces the image to fit the cover, so lift a detail from Albert Bierstadt's "Among the Sierra Nevada Mountains California" rather than condense the entire painting. Also, avoid picketers' signs and newspaper headlines because the words in the image will compete for the viewer's attention with your title and name.

If your editor invites your ideas, propose four to six options. When you agree upon one of them, ask who will take responsibility for obtaining the image and negotiating with the copyright holder or the owner of the physical image. If these responsibilities fall to you, make sure you know exactly which image format (.jpeg, .tiff) and what density (dpi) the designers require. Also inquire what permissions to request: sales territory, digital as well as print editions, promotional uses, and duration of the permission term. Ideally, you will be able to obtain permission to use the image on all current and future editions, both digital and print, for sale throughout the world and for all related advertising. You're requesting nonexclusive permission, meaning that the copyright holder may grant the same permissions package to anyone else. On the permission form, note any restrictions on cropping, wrapping, or overprinting the image. Designers can usually work around these prohibitions; but they prefer the liberty to trim the image, pull it around to the back cover, or place lettering over it if they are so inspired. Typically, the more generous the permissions package, the higher the cost; don't hesitate to ask what portion the publisher will pay.

By thinking like a designer, you improve the likelihood that the publishing team will agree to use your proposed image. Even if your editor does not solicit your ideas, with an understanding of the underlying logic, you can better evaluate the cover created for your book and offer suggestions in terms meaningful to the designers.

Title

You invented a working title when you assembled your book proposal. Different publishers have different timelines, so you and your editor may have finalized the title when you signed the contract. If not, the time has now come. Does that witty title tantalize you again? Please, suppress the temptation! Remember that not everyone shares your taste, education, or cultural background. What you consider witty might bewilder or offend. Furthermore, potential readers will discover your book primarily through a keyword search. Unless your title includes those words, their search won't capture your book. If by chance they do call up your book, they won't pause to decipher your incomprehensible title in the list of more communicative ones.

Along the way, your editor and the reviewers proposed alternative titles. One of these, with a few adjustments, could be just what you want. If not, repeat the title development process introduced in a previous chapter. Skim your table of contents and introduction, circling all the keywords. Then test different forms (i.e., nouns, verbs, adjectives) and arrangements of these words until you arrive at the optimal combination: a main title broadly announcing your topic and a subtitle clarifying that topic, which together point to your argument. In the event that your marketing team isn't quite satisfied, listen to their suggestions. Think of them as your hired guns, commissioned to nab unsuspecting readers and ring up book sales.

You can prevent confusion by placing your preferred title on the title page of your manuscript as well as drawing attention to it in your submission cover letter. As an extra precaution, use the "find" feature in your writing program to search your entire manuscript and eliminate former incarnations of the title. It's embarrassing—and embarrassingly common—for a discarded book title or chapter title to pop up in a published book, causing the reader either puzzlement or amusement.

Author's Name

Why am I reminding you to specify how you want your name to appear on your book? Oh, the stories I could tell . . . . And let me add that this is the time to resolve any quarrels among your co-authors regarding the order of your names.

FRONT MATTER

Let's click through your front matter, from your title page to the first page of your introduction.[2]

Title Page

In your manuscript, the title page should include your main title, subtitle, and name. Your copyeditor will fill in your publisher's name and city. He will also add the half title page (i.e., the page with only the main title), the copyright page, and the series page (if applicable).

Frontispiece

A frontispiece serves as an easy-to-find visual reference as the reader progresses through your book. For example, a portrait of Charles Darwin works well in a biography; a map, in an analysis of military tactics in the Spanish Civil War; a family tree, in a study of the Borgia dynasty.

Dedication

A heartfelt dedication is a lovely thing. But note that rewriting your will is easier than rewriting your dedication page and choose your dedicatee with care. Your publisher can't stop the presses when you get divorced or your cat shreds your antique rug.

Epigraph

As I was finishing this manuscript, I discovered the perfect epigraph: "I cannot offer You a greater gift than the prospect of Your understanding in the shortest period all that I have experienced and learned over so many years and with so much danger and hardship."[3] This line, with which Machiavelli dedicated *The Prince* to Lorenzo de' Medici, encapsulates the reason for writing this guide. But the man who astutely marked the distinction between virtue and *virtù* has a mixed reputation, the result of careless misreadings and deliberate misunderstandings over the past 500 years. In the end, I chose not to include this quotation because an epigraph says as much about the author as about the book. Consider how a quotation from Winnie the Pooh or Dr. Seuss will impress readers. Wouldn't you rather align yourself with Benjamin Franklin, Antonio Gramsci, or Albert Einstein?

Table of Contents

When you finish typing your table of contents, read the chapter titles once more. Does each one accurately describe the subject of the chapter? Taken together, do they convey a sense of movement through your storyline or argument? If you fine-tune them, be sure to enter the changes throughout your manuscript. At this point, you don't need to include page numbers on your table of contents. Later, the copyeditor will ensure that it corresponds with the pages in your published book.

Lists of Illustrations, Tables, and Figures

Depending on the conventions of your field and the number and variety of features in your book, you may not need a list of illustrations, a list of tables, or a list of figures (e.g., maps, graphics). Nonetheless, I suggest that you include them with your final manuscript because, whether or not they will appear in your book, they serve as inventories for both you and your publishing team. As with the table of contents, you don't need to worry about page numbers, but you want to check the numbering for each feature (e.g., Table 1.1, Table 1.2, Table 2.1). Also as with the table of contents, reading these lists will help you spot infelicities and smooth the logical flow. If you make any adjustments, please take a moment to update the illustration captions and table labels as well as the callouts in your manuscript, which indicate where each item should appear in your book.

Foreword

A foreword, literately a word that comes at the fore, serves the same function as the introduction for a guest speaker. The host highlights the speaker's credentials and the significance of the forthcoming presentation and then hands him the microphone. Likewise, in a book, a series editor or an authority on the topic sets the stage for the author. Typically, the publisher makes the decision regarding a foreword and takes responsibility for commissioning it; but don't hesitate to suggest the names of individuals who might write it, especially if they have worked with your publisher in the past or if you know them personally. You want to have this conversation with your acquisitions editor as soon as possible because it may take a few months for the writer to complete the foreword. Also, no matter who the esteemed writer may be, ask to read it before your copyeditor begins working on it.

Preface

Whereas a foreword is written by someone else, the preface is written by the author. Continuing the previous analogy, when the guest speaker takes the stage, he warms the audience with a few words about himself. If you choose to write a preface, you might think of it as a personal note to the reader. Tell a brief story about how the topic grabbed your interest, what inspired you to write your book, and what you hope to achieve. Readers don't want a confessional; they want a brief self-introduction from the expert with whom they will spend many hours.

Acknowledgments

Your preface leads directly to your acknowledgments. In fact, many authors combine the two into a section titled "Preface and Acknowledgments." Definitely mention institutions that provided funding, sources, or other support. Instead of enumerating all the individuals who helped you along the way, it's better to list generic categories. Thank interview subjects, reviewers, colleagues, conference panel participants, research assistants, librarians, archivists, and your publishing team. Reserve personal thanks for those few who gave unstinting and indispensable assistance. Note that acknowledgments occasionally appear in the back matter rather than the front matter.

List of Abbreviations

Most readers are familiar with abbreviations for major U.S. government agencies and world organizations that appear in local newspapers every day, but you cannot assume that because they equate *UN* with the United Nations, they will recognize *WTO* as the World Trade Organization. If you intend to abbreviate a name, title, or concept, follow it immediately by the abbreviation in parentheses the first time you use it in each chapter, like so: International Criminal Court (ICC). Then, throughout the rest of the chapter, use only the abbreviation. When you mention the same person, entity, or concept in the next chapter, begin again with the full term and the abbreviation. If you use more than a dozen abbreviations or acronyms, a list placed either at the front or the back of your book will provide a handy reference, particularly for readers from outside your specialty.

Chronology

As with the list of abbreviations, a chronology may appear in the front matter or the back matter. Your readers will appreciate this reference if your narrative follows a sequence of events or your argument focuses on particular historic moments. Select the range and the individual points on the chronology according to your narrative and argument. That is, your chronology should not only help readers to keep track of your storyline but also supplement your evidence.

BACK MATTER

Do you remember riding the school bus? The well-behaved elementary students sat up front while the unruly junior high kids ranged about in the rear.

As eager as you are to get your manuscript off your desk, don't let it resemble the bus. Although certain parts of your book appear at the end, you don't want to treat them as afterthoughts. Done right, your appendices, notes, and list of references substantiate your argument.

Appendices

Appendices provide space for ancillary material relevant to your book as a whole and not readily available elsewhere. Consider both these factors as you determine what to include. Information that the reader will want to consult at various points in your book belongs in an appendix. But survey questions, data graphics, and text tables specific to a chapter should stay within the perimeter of that chapter. Believe it or not, readers would rather Google a list of U.S. presidents or Beethoven's works for solo piano than pay for and tout around page after page of common information. Please don't treat appendices as storage bins for stuff that seems important but has no proper place. Two appendices, or possibly three depending on your field, should suffice. More suggest pedantry or indecisiveness.

Endnotes

In order to save labor and production costs, most scholarly publishers prefer endnotes, gathered either at the end of each chapter or, more commonly, at the back of the book. You may succeed in convincing your editor to make an exception for your book, however, if footnotes are conventional in your field. Regardless of the placement, you can win points with your publishing team by making sure your notes are complete, down to the last jot and tittle of each citation.

Glossary

Do you work in a profession such as architecture or a field of study such as dendrochronology that employs unique words or deploys common words in unique ways? If so, you should consider creating a glossary. But as with the other reference materials I've discussed, the benefits of a glossary should outweigh the costs to your readers.

List of References

What is the distinction between a bibliography and a list of references? A cynical copyeditor would answer that a bibliography enumerates every potential source that an author touched, sniffed, or glimpsed from afar in the course of his research, whereas a list of references includes only those sources that he actually uses and cites. Not surprisingly, most publishers prefer a list of references. Your reader will thank you for consolidating the list as much as possible. Go ahead and combine different categories (e.g., primary sources, secondary sources) and genres (e.g., articles, books, recordings) into a single list, sorted alphabetically. You could provide separate lists for distinct categories, particularly if they highlight the originality of your research, such as "Archival Collections" and "Interviews Conducted by the Author." Place these directly in front of your standard list of references. Also, consider adding a list of abbreviations exclusively for your references so that you can condense your citations.

Index

Here's the good news: you won't create your index until the designers transform your manuscript into numbered book pages. When the time comes, I recommend that you hire an indexer because even the best software cannot replace the skills of an experienced professional. Your production manager can provide a list of freelancers with whom he has previously worked and who, consequently, are familiar with your publisher's requirements. If you are considering creating the index yourself, take a look at the chapter devoted to indexes in *The Chicago Manual of Style*. If, after reading these guidelines, you remain undaunted, you will find more explicit instructions in Nancy C. Mulvany's *Indexing Books* and resources on the American Society for Indexing's website.[4]

AIDS FOR YOUR PUBLISHING TEAM

I've covered all the standard parts of a scholarly book, but by submitting a few additional documents with your final manuscript, you'll win extra points with your team and expedite the publishing process.

List of Figures or List of Illustrations with Captions

You already did half the work when you typed the list of figures and/or the list of illustrations for the front matter. Now copy those lists and insert the caption for each item. To review, a caption should answer the following questions: 1) What is the reader looking at? 2) Why does it matter? 3) Where does it come from? Describe the image. Note its relevance to your topic or argument. Then state the source of the image or the data, using the precise wording specified by the individual or organization that gave you permission to use it. By making sure that all the items are properly numbered and providing complete information for each, you prevent a great deal of confusion.

Package of Permission Forms

Once you have your lists with captions, you can efficiently package your permission forms. I suggest that you write on each one the number of the figure or illustration to which it corresponds and then scan the forms in numerical order and gather them into a single PDF labeled with your last name and the word "permissions."

Stylesheet

A stylesheet serves as the master guide for spelling, punctuation, and terminology. Do you use any non-English words (e.g., *gesundheit, boca abajo, savoir faire*)? Technical terms (e.g., braze interface, man-year, barding)? Proper nouns with uncommon spellings (e.g., Lynda, Jonesboro, Pittsburgh)? Descriptive terms preferred by particular ethnic, racial, or minority groups (e.g., Black or African American, Native American or American Indian)? If you didn't compile a stylesheet as you were writing your manuscript, please do so now as you give it a final once-over in order to ensure that the copyeditor follows your preferences consistently throughout your book.

Contact Information

Finally, you want to make sure that your publishing team knows how to contact you by phone and by e-mail as well as where to ship packages. Notify them that you're spending the summer at your cabin in northern Minnesota, you're expecting a baby in mid-February, you're going to New Zealand in three weeks for a family wedding. The better you keep your team informed, the more willingly they'll attempt to coordinate the production schedule with your own calendar.

FROM YOUR DESK TO THE PRESS

Congratulations! Your manuscript and all the fixings are ready for submission. When you e-mail the package to your acquisitions editor, please remember to use your standard subject line: your last name / your title. This time, though, after your title you can add in capital letters "FINAL MANU-SCRIPT." Doesn't typing those words feel good? In the short message accompanying your submission, clearly state your title and the way you want your name to appear on your book. Provide an inventory of the attachments, and please double-check that you did attach each one.

Your efforts to prepare your manuscript according to your publisher's guidelines and to ensure that every piece is in order will go a long way to establish a collegial relationship with your publishing team. They should reciprocate with the same courtesy and respect as, together, you work to produce a handsome, high-quality book. In the next chapter, I discuss the publishing process as well as the third phase of your charm offensive: the marketing campaign.

NOTES

1. Metropolitan Museum of Art, Open Access for Scholarly Content Policy, accessed 28 October 2016, http://www.metmuseum.org/about-the-met /policies-and-documents/image-resources.
2. You'll find more information about the parts of the book and their order of appearance in the publishers' bible, University of Chicago Press Staff, eds., *The Chicago Manual of Style: The Essential Guide for Writers, Editors, and Publishers*, 16th ed. (Chicago: University of Chicago Press, 2010). My presentation here generally follows the order specified by the manual, although for the sake of expediency I subsume the half title and title pages under a single subhead. Your publisher will use the proper arrangement: half title, series title/frontispiece/blank page, title page, copyright page.
3. Niccoló Machiavelli, *The Essential Writings of Machiavelli*, trans. and ed. Peter Constantine (New York: Random House, Inc., 2007), 5.
4. University of Chicago Press Staff, eds., *Indexes: A Chapter from The Chicago Manual of Style Sixteenth Edition* (Chicago: University of Chicago Press, 2010); Nancy C. Mulvany, *Indexing Books*, 2nd ed. (Chicago: University of Chicago Press, 2005); American Society for Indexing, accessed 10 March 2017, https://www.asindexing.org/.

CHAPTER 10 _____

Publishing and Marketing Your Book: A Member of the Team

The FedEx driver demanded that I show him my ID and sign the receipt before he turned over the package. I'd started my apprenticeship with the history editor at Johns Hopkins University Press only the previous week, and I couldn't guess why a professor at a prestigious New England university had sent me this huge box. Inside I found a roll of bubble wrap with a smaller box, sealed with shipping tape, at its core. And inside that box, underneath another layer of bubble wrap, rested a 3-inch stack of paper and a pair of floppy disks.

For the FedEx carrier, the package's content was a just pile of paper. For me, it was a raw manuscript. For the author, however, it represented nearly a decade of research, years of writing, and a soon-to-be book that could secure or topple his reputation. Now he was entrusting his book to strangers: the editorial assistants, copyeditors, production managers, designers, marketers, printers, and warehouse workers on Hopkins's publishing team. He had to trust our skill and goodwill. No amount of packing tape and bubble wrap would protect his book. And, in a sense, he was counting on us to fortify it against reviewers' charges of inaccuracy, sloppiness, and irrelevance. Before I hauled the packaging to the dumpster, I taped a 2"×2" square of bubble wrap to the ledge of my cubicle shelf as a reminder of every author's anxieties. Two decades have passed since that morning. Go ask authors who've worked with me whether or not I remembered.

In this chapter, I provide an overview of the publishing process and empower you as an author to work alongside your team. Although the details and the insider terminology vary somewhat from publisher to publisher, I explain what to expect from your teammates and, equally important, what they expect of you. Every team member is a professional, and each has a personal ambition beyond the collective goal of producing a successful book. Your acquisitions editor covets the leading book prizes in your field, whereas your designer may want to fashion an award-winning cover. Your marketers may seek an interview on the local NPR station, a review in the *Wall Street Journal*, or a favorable mention in *The Huffington Post*.

Preparing your manuscript carefully goes a long way toward winning your teammates' goodwill because, generally speaking, publishing professionals appreciate order, precision, efficiency, and predictability. They prefer a form to an epistle. Why? Reducing ambiguity reduces the potential for mistakes and misunderstandings. Moreover, expediting routine matters leaves more energy for thought and creativity.

As you meet your publishing team, you'll discover the surprisingly different ways in which others understand your book, and this discovery will enable you to market it more effectively to diverse audiences. For this reason, after swiftly covering the publishing process, I devote the rest of this chapter to marketing. I suggest how to coordinate efforts with your marketing team and employ nontraditional techniques for promoting your book.

THE LAUNCH

When you visit a publishing house, you won't hear many nautical terms, but you may witness directors and key staff representing every department hastening to a *launch*. Actually, as you'll discover if you tag along, the term appears to be derived more immediately from astronautics. Just as the ground crew gathers to inspect the ship before launching it into space, during that meeting production managers, designers, marketers, and occasionally the press director himself gather to inspect the manuscript, which transformed into a book, they will launch into the world. The acquisitions editor plots the intended course. Then, each speaking from his own specialized perspective, her colleagues recommend technical adjustments, identify challenges, and discuss strategies. With the particulars in order and the crew in place, the production countdown begins.

Although the marketers don't take over until the publication blastoff, the team largely defers to them because they pilot the book in flight, steering it through advertising outlets, review media, and retail websites, into readers' hands and onto their screens. Just as planetary orbits and local weather patterns guide the choice of the blastoff date for a space mission, predictable cycles and anticipated events guide the choice of a book's publication date. Marketers base their calculations on conference schedules, election cycles, and the rhythms of the academic year, as well as international political events and the anniversaries of relevant institutions, individuals, and historic events. Marketers also streamline the title, tune the advertising description, and spark up the cover design so that the book can break through the atmosphere of indifference. Once beyond this barrier, the intrinsic interest of the topic, joined with the author's propulsive argument and writing style, should build sales momentum.

After your team holds the launch meeting, you can expect a few additional requests such as slight alterations to a chapter title or an adjustment to your tables. Then your acquisitions editor transmits your manuscript to the production manager. Just as his title designates, this individual coordinates every step of the process as your manuscript passes from one department to the next. Here I should mention that most publishers hire freelance copyeditors and designers. Occasionally, a publisher employs a separate company, known as a book packager, to handle the entire production process. Regardless of whether you work with regular staff, freelancers, or some combination, you should expect an in-house production manager to take charge of your manuscript and serve as your primary contact. And because you took the time to investigate this publisher before signing your book contract, you should feel confident that you partnered with a team of professionals.

COPYEDITING

Whatever your hopes or fears, your copyeditor does not revise your prose. Rather, she reads your manuscript with two questions in mind: Does it conform to conventional standards? Does it make sense? Following the stylesheet you provided, your publisher's stylesheet (aka house style), and the revered *Chicago Manual of Style* or the style guide favored by your field, she standardizes grammar, punctuation, and terminology. On the surface, her job appears mechanical, a task that artificial intelligence will assume some day soon. But in fact, doing this task properly requires close attention to and a basic understanding of the text. Aware that she is likely the first person outside your field to encounter your manuscript, your copyeditor highlights esoteric jargon, points to awkward phrases, and generally asks for clarification of potentially confusing passages, all in order to make your book more comprehensible to nonspecialist readers and fellow experts alike.

When your copyeditor finishes, you receive a digital file of your manuscript showing the changes. (Publishers in the United States typically use software that tracks changes.) You then have a specified period of time to review and respond to her queries and recommended corrections. Resist the urge to rewrite your manuscript! I've known authors who, upon receiving the copyedited manuscript, suffered a sudden attack of self-doubt, decided the whole project was worthless, and wanted to start over. Have confidence! You're an expert. This manuscript represents your best work. In addition, it passed multiple levels of rigorous review and underwent multiple rounds of revision. At this point, you need only double-check the copyeditor's alterations and approve those recommendations that you believe will improve your manuscript. Yes, you can reject those that you don't agree with. But if you find yourself undoing a large percentage of your copyeditor's work, you want to talk with your production manager because occasionally the press assigns a manuscript to a copyeditor who is overzealous or unfamiliar with your field.

By returning your manuscript on or before the requested due date, you help to keep your book on schedule. I will spare you the woeful tales of authors who repeatedly missed due dates and, consequently, delayed publication by as much as 10 or 14 months. Your team counts on you!

INTERIOR DESIGN

After you return your manuscript and your copyeditor confirms the agreed-upon changes, your production manager hands it to a designer. A book's cover may get more attention, but the designer devotes more effort to its insides. A good interior design conducts the reader's eyes across the pages, enabling the efficient transport of information from the author to the reader. In your own reading, if you've ever found yourself trapped by walls of solid text or wearied by a random clutter of graphs, you know that the interior design determines comprehensibility as well as approachability. You've already given thought to the placement of the various elements of your book. And you considered the reader's experience as you made decisions about the length of paragraphs, the size of block quotations, the complexity of tables, the appearance of graphics, and the appeal of photographs. Now, working from your preliminary layout, your designer chooses the fonts, sets the margins, and adds running heads as he transforms manuscript pages into book pages. In the jargon of the publishing industry, these are the page proofs.

Just as you reviewed the copyedited manuscript, you review the page proofs. At this point, you need only scan for typographical errors. Of course, you want to catch any misspellings, and you should pay particular attention to numbers (e.g., dates, statistics, numeric tables). Your production manager proofreads the front matter and the chapter titles, but don't rely on him to proofread your entire book. Once again, you may feel an urge to rewrite. If you need to fortify your resistance, take a look at your publishing contract. Many publishers, anticipating an author's self-doubt, charge for extensive changes to the page proofs.

INDEX

While you have your contract in hand, check the clause regarding the index. If you are responsible for providing the index, this is the time to create it. Thus far you've entrusted your book only to professionals, experts like yourself but who specialize in specific aspects of publishing. So please, for the sake of your own reputation, I urge you to hire an experienced indexer. Your production manager can refer you to freelancers who regularly work with your team and who therefore know the requirements. If you want to try creating the index yourself, I recommend following the guidelines in *The Chicago Manual of Style* and Nancy C. Mulvany's *Indexing Books*. The American Society for Indexing also lists resources on its website.[1] As before, by returning your page proofs with the index on or before the due date, you help to keep your book on schedule.

COVER DESIGN

Because readers and reviewers actually do base their initial judgment on a book's cover, your designer carefully considers every element, from the image and color scheme to the font and the alignment of the type. Sometimes he creates three or four versions, called mockups, which he shares with fellow designers, the acquisitions editor, the marketing director, and the press director. Following the group's suggestions, he perfects the preferred version, which the production manager or the acquisitions editor then forwards to the author.

A word of warning: even if you helped to select the image, your book cover will not look exactly like you envisioned it. You may love it. You may hate it. You may want to alter it just a bit. Whatever your reaction, try to be open-minded. Wait a day before you phone or e-mail your acquisitions editor. A publisher reserves the right of final approval to guard against the author whose imagination far outruns his experience with graphic design and whose ideas could jeopardize sales. I knew an author who demanded the color orange, although it was completely inappropriate for his subject, because he wanted his book to stand out on bookstore shelves. One author insisted on a black-and-white mug shot of an unknown historic figure on his cover, whereas another wanted an artistically distorted photograph of a government building. None of these ideas were wrong, per se; they simply did not align with the recommendations of experienced designers and marketers. Fortunately, these authors were reasonable, congenial individuals, willing to acknowledge the expertise of others and ready to accept a compromise. And the author who objected to a smoking cigarette on his book cover because it inadvertently promoted tobacco products persuaded the team to find a different image.

So, when you first see the cover design, remind yourself that your team wants your book to succeed as much as you do. If you have serious concerns, by all means put on your most charming self and alert your acquisitions editor. But I'm betting that, because you and she discussed your cover long before this point, you will be delighted.

PRINTING, WAREHOUSING, AND DISTRIBUTION

After you return the page proofs and index, your production manager dispatches the digital files to the printer, who manufactures the physical books and ships them to the warehouse. Since the industry adopted the Just-in-Time (JIT) approach, publishers have reduced the initial print run. Rather than print two or three years' worth of stock, they now print enough copies to cover the sales estimate for the first six months. As this supply dwindles, they print more copies to fill orders and to keep a handful of books on reserve in case of a sudden demand. Many small and medium-sized publishers contract with larger publishers or book companies for warehousing and distribution. The Chicago Distribution Center (CDC), for example, provides its clients with warehousing, order fulfillment, and a menu of other services related to marketing and sales. Whereas the CDC, operated by the University of Chicago Press, caters to scholarly publishers, the National Book Network (NBN) and similar companies focus on commercial publishers. Because digital formats (e.g., Kindle, Nook, iBook, PDF, ePub) and sales channels are so varied and complex, smaller publishers commonly contract for the creation and distribution of e-books. In addition, a publisher may participate in a program such as ProQuest's ebrary, which aggregates e-books from many publishers into subscription packages expressly designed for academic libraries.[2]

In short, your marketing and distribution team spans the globe and permeates a range of markets, extending your book's reach not only across bookstore shelves and retail websites on all continents, but also deep into databases and library catalogs. And this reach greatly augments the possibility that your book will land in readers' hands or on their screens.

MARKETING

A marketing campaign for a book resembles a political campaign in many ways. You must target the right audiences, with the right message, through the right media in order to get their attention, convince them of your cause, and move them to action. As an author, you want to project an authoritative yet appealing persona and persuade your audiences to purchase your book.

Just as a candidate seeks the endorsement of respected public figures, you seek the endorsement of prominent reviewers and prize committees. In the following pages, I describe ways that you can coordinate efforts with your team for a successful book campaign.

I also discuss strategies for advancing your career. As I've reiterated throughout this guide, publishing a book serves the larger goal of establishing yourself as *the* go-to expert on your topic. If you're seriously committed to this goal, I urge you to look at the literature on personal branding, a concept that originated in the business world. When branding proved successful for corporations, executives adapted these techniques to package and promote themselves. You, too, can employ personal branding strategies, whether you're an organic farmer, a journalist, a minister, a philosopher, a social worker, or a mathematics professor. To get started, I recommend *Personal Branding for Dummies*, 2nd edition, by Susan Chritton.[3]

Behind-the-scenes promotional efforts begin several months before publication, as your marketing team collects endorsements (aka blurbs) from well-known individuals for your book cover and webpage. During this time, your team also markets the subsidiary rights, especially translation rights, but also rights for various forms of adaptation, which you granted your publisher when you signed the contract. Through international book fairs, trade journals, and the efforts of in-house staff or hired agents, your team alerts publishers, retailers, and the media around the world to your forthcoming book. The marketing campaign really takes off in the weeks following publication. It remains intense for about six months, or one publishing season, before gradually fading toward the end of the first year, at which point your book moves from the frontlist to the backlist of older titles. Although scholarly publishers continuously promote all their books, they understandably allocate most of their marketing budget to frontlist titles.

Targeting Members of Professional Societies

Remember the research you did when you were looking for a publisher? Now you can repurpose that information for your marketing campaign. Soon after the launch meeting, you received a marketing information form asking for a list of relevant journals that review books, conferences that host book exhibits, and prizes for which your book is eligible. These three venues bring your book to the attention of professional societies in your field and neighboring fields where you're likely to find most of your readers. Although your marketing team handles review copies, conference displays, and prize submissions, you can amplify their efforts. You should network regularly with editorial board members, especially book review editors. As your book nears publication, find a reason to mention it in an e-mail message or an in-person conversation and offer to provide a list of individuals who might review it for the journal. In the same way, find out who serves on prize committees and, if they are not already in your network, reach out to them.

Don't hesitate to make a "cold call" or send a "cold e-mail." Contacting someone whom you don't know to solicit information or merely to introduce oneself is common practice in the business world. If you're uncertain how to begin, I suggest opening your message with praise for the person's work—a presentation, an article, a blog—and explaining how it inspires or challenges your own work. Then describe your forthcoming book and your goals for it. You could conclude by expressing the hope that this person will find your book of interest. Sincerity and brevity are essential for a successful cold call. Your compliments must be genuine, and you must not write more than six or eight lines.

During the 18 months following the publication of your book, you want to give as many conference presentations as possible. For major conferences in your field, organize a book roundtable and invite prominent individuals. If they plan to attend the conference anyway, they may welcome the opportunity to participate in a lively discussion with their peers about a brand-new book. Don't be discouraged if an international association doesn't grant you space on the conference agenda. You may have more success if you propose a round-table for a meeting of the association's regional or special-interest divisions. And you can always present an individual paper. For conferences in neighbor-ing fields, you might arrange a panel with researchers who work on related topics. Here again, you should network with such individuals on a regular basis; but if you haven't yet done so, it's not too late to start cold calling. Ideally, you want to assemble a panel that demonstrates that although you approach the topic from a somewhat different angle, you contribute to the essential ques-tions and most pressing debates in this particular field. You have expertise that this audience needs! They simply won't know it until they see your pre-sentation, delivered with your ceaseless charm and confident authority.

In order to coordinate efforts, keep your marketing team informed about your conference presentations. Scholarly publishers rarely pay for a book tour or other travel related to book promotion, so please talk with your marketing team before you submit your airfare and hotel bills under the mistaken assumption that you will receive reimbursement. But don't hesitate to ask for a professionally designed book flyer with the publisher's logo, and preferably with a discount code, which you can distribute by e-mail and in person. And if your publisher plans to host a conference exhibit, offer to spend an hour or two at the booth for a book signing.

You can also reach members of professional societies, both in your field and in neighboring fields, by publishing a journal article or two within four months of your book's publication date. Talk with your acquisitions editor about adapting selections from your manuscript. If this option is not feasible, you can still publish articles related to your topic. And don't forget to mention your forthcoming book in the author tagline!

Targeting Online Audiences

In recent years, publishers have reallocated marketing dollars from mass mailings, print ads, and other forms of print advertising to e-mail blasts, social media, and creative uses of a variety of digital forums. They've also invested heavily in their websites, although my colleagues at university presses report that the majority of online visitors come to learn about books before purchasing them from an online retailer. I suspect the same is true for commercial publishers. Please note that your publisher has no direct control over an online retailer. Your team provides information, called metadata, which includes everything from your name, your book's title, and the ISBN number to the page count, the number of illustrations, and your book's dimensions. Information can slip in the transit from the publisher's database to the retailer's, and making corrections takes time. For this reason, in the months preceding publication of your book, you might periodically check the most prominent retail sites and alert your marketing team as soon as you spot an error.

Amazon offers a feature called Author Central. When you create an account, you can upload your photo, videos, information about yourself, and a calendar of your presentations and book signings. You can also monitor your book sales. At the moment I'm writing this chapter, Amazon does not charge for this service.[4] If this feature becomes popular and profitable, other retailers may provide something similar.

Your publisher may post selections from the front matter and text with the aim of enticing web visitors to purchase your book. Before you repost these selections or others, please read your contract and consult your marketing team. There are strategic reasons for making particular sections available in particular venues. Scattering a page here, a chapter there, a table here, a graph there across the web will not improve your chances of making a sale. Likewise, if you want to use your book for teaching, please check your contract. Your publisher may grant you an author's dispensation. Then again, you may need to treat your own book as you would any other copyrighted material that you use in your classes.

Just as you supplement your team's efforts to reach professional societies, you can support the online marketing campaign by contributing to the press blog and joining the conversation on social media. But don't set up camp just yet. Advance the front. Participate in blogs relevant to your topic by composing your own posts and commenting on others. Join listserv conversations and, as appropriate, mention your book. Even if you don't blatantly advertise, you demonstrate that you're both an expert on the topic and a writer whom readers want to engage with.

If you have a website and social media accounts, update them. If you don't, create them now.[5] Incorporate videos, photographs, and audio clips—whatever features enliven flat text. Be warned, however, that keeping your online presence fresh demands an extraordinary amount of time. On your calendar, block out several hours for social media each week during the two months following your publication date and additional hours in the days surrounding any of your live or virtual presentations. When the rush of your book campaign tapers off, continue using these networking and communication tools regularly to stay visible. It's better to maintain a central website and one or two social media accounts with weekly, if not daily, activity than to leave behind a digital ghost town with your name on every neglected building. And, as always, ask your marketing team about their preferred social media and ways to coordinate efforts.

Targeting Local Audiences

Finally, no matter what your topic, you'll find marketing opportunities close to home. Speak to a senior citizens' group. Pay a visit to the Boy Scouts. Give a reading at your public library. Offer a book signing at your hometown bookstore. Although you may fear that your neighbors have no interest in the popular election of representatives for the European Parliament, by this time you've had enough experience with answering the question: "So what is your book about anyway?" from relatives, tennis buddies, and fellow passengers that you can identify a point of interest for any audience. If you get stumped, look for a hook in the news or in a typical person's daily experience. Journalists excel at transforming seemingly abstruse topics into compelling and relevant stories. You can do the same. Speaking of journalists, make sure that you appear on the list of experts posted by your university's public relations office. Declare your willingness to answer calls from reporters. You might write an editorial for your local newspaper or arrange an interview with the local radio station.

You'll find many promotional opportunities in your professional neighborhood as well. Volunteer to speak at your department's colloquium, address graduate seminars, and visit undergraduate classes. Contact colleagues at other universities and offer to give presentations on their campuses. If travel expenses concern you, propose live webinars. Flipping through your e-mail address book, you'll remember the SPCA representative, the curator at the

U.S. Holocaust Museum, the researcher at Carnegie Mellon's Robotics Institute. Would these individuals welcome you as a speaker? You won't know until you ask. I urge you to reach out to individuals who share your interests, especially if they contributed to your research in some way.

CAMPAIGNING BOLDLY

Do you still feel shy about promoting your book? In an interview with Rachel Toor for *The Chronicle of Higher Education*, Niko Pfund, president of Oxford University Press, observes: "Readers love to hear from, and connect with, authors."[6] Whereas he proceeds to describe Oxford's use of social media, his observation applies equally to speaking events, virtual presentations, radio interviews, and newspaper editorials. By spreading your campaign through many venues, you are not merely pitching your book, you are also offering readers an opportunity to engage with you, the expert author, beyond the page.

You're not convinced yet? Okay, I'll tell you what I told the author who phoned me late one afternoon to ask about the etiquette of submitting his book for a prestigious prize. This author is the most courteous, most self-effacing person I know. Anything even faintly resembling self-promotion makes him cringe. After 15 minutes of assuring him that this kind of self-promotion is entirely acceptable, I finally reminded him that the entire publishing team contributed to his book. By submitting it for this prize, he was acknowledging the excellent work of the production manager, the copyeditor, and the indexer; the designers and marketers; the anonymous reviewers and the innumerable colleagues with whom he discussed his project. "You may not believe you deserve the prize," I said, "but we deserve the chance to compete for it. And when you win, you can accept it on behalf of us all."

Now I'm telling you. If you want to recognize your publishing team, get out there and proudly promote your book.

NOTES

1. University of Chicago Press Staff, eds., *Indexes: A Chapter from The Chicago Manual of Style*, 16th ed. (Chicago: University of Chicago Press, 2010); Nancy C. Mulvany, *Indexing Books*, 2nd ed. (Chicago: University of Chicago Press, 2005); American Society for Indexing, accessed 10 March 2017, https://www.asindexing.org/.
2. Chicago Distribution Center, accessed 20 November 2016, http://press .uchicago.edu/cdc.html; National Book Network, accessed 20 November

2016, http://nbnbooks.com/; ebrary, accessed 20 November 2016, http://proquest.libguides.com/ebrary.

3. Susan Chritton, *Personal Branding for Dummies*, 2nd ed. (Hoboken, New Jersey: John Wiley & Sons, Inc., 2014).

4. Amazon Author Central, accessed 23 November 2016, https://authorcentral.amazon.com/.

5. On their website, the marketing professionals at the University of California Press host an Author Toolkit, complete with video tutorials, in which they explain how to take advantage of social media and why you should do so. They make this toolkit freely available on the website, and any author can benefit from their advice; accessed 23 November 2016, http://www.ucpress.edu/toolkit.php. Oxford University Press lists the leading platforms and suggests how to use them in the Social Media Author Guidelines, accessed 23 November 2016, https://global.oup.com/academic/authors/social-media-guidelines.

6. Rachel Toor, "Write a Book and Become an Employee of Your Former Self," *The Chronicle of Higher Education* 13 October 2014, accessed 24 November 2016, http://www.chronicle.com/article/Write-a-BookBecome-an/149363.

Conclusion: Giving Back

Congratulations! By publishing your book, you've established yourself as the go-to expert on your topic. Your book will open new opportunities for career advancement and further research. It will also open new opportunities to give back to your professional community because it adds a lustrous new layer to your credentials: you're now a published author. As your book gains recognition, anticipate more requests to review manuscripts and invitations to serve on committees. Most likely you expect and look forward to these opportunities.

You may be surprised, however, when a colleague consults you about how to publish his own book or the graduate students ask you to talk about publishing at their monthly brown bag. Granted, you aren't an expert on publishing a book; you've only done it once. But you have done it once and done it successfully. Share what you know. When you meet a young scholar, tell her about a press that you discovered during your survey of the publishing landscape. Offer your struggling officemate a few tips on how to stay motivated during the long process of writing. Show a colleague how to respond to the criticism of peer reviewers. And if you found this guide useful, as I hope you have, please pass it on.

References

Amazon Author Central. Accessed 23 November 2016. https://authorcentral .amazon.com/.

American Musicological Society. "Best Practices in the Fair Use of Copyrighted Materials and Music Scholarship" (2010). Accessed 8 March 2017. http://www.ams-net.org/AMS_Fair_Use_Statement.pdf.

American Society for Indexing. Accessed 10 March 2017. https://www .asindexing.org/.

Amherst College Press. Accessed 27 December 2016. https://acpress.amherst .edu/about/.

"An Old Lady Who Swallowed a Fly." The British Council, 2012. Accessed 29 November 2016. www.britishcouncil.org/learnenglishkids.

Association of American University Presses. Accessed 27 December 2016. http://www.aaupnet.org/about-aaup/about-university-presses.

Association of American University Presses, Acquisitions Editorial Committee. *Best Practices in Peer Review*. New York: Association of American University Presses, 2016. Accessed 30 December 2016. http://www.aaupnet.org /policy-areas/peer-review.

Association of American University Presses. "Subject Area Grid." Accessed 24 February 2017. http://www.aaupnet.org/images/stories/documents/aau psubjectgrid2017.pdf.

Bedford, Daniel, and John Cook. *Climate Change: Examining the Facts*. Santa Barbara, CA: ABC-CLIO, LLC, 2016.

Booth, Wayne C., Gregory G. Colomb, Joseph M. Williams, Joseph Bizup, and William T. Fitzgerald. *The Craft of Research*. 4th ed. Chicago: University of Chicago Press, 2016.

Brewer, Robert Lee, ed. *Writer's Market 2017: The Most Trusted Guide to Getting Published.* New York: F+W Media, 2016. Writers Market, Category Search—Book Publishers. Accessed 24 February 2017. http://www.writers market.com/MarketListings/BookPublishers/search.

Carroll, Lewis. *Through the Looking Glass, and What Alice Found There.* Illustrated by John Tenniel. n.p.: Thomas Y. Crowell & Co., 1893. Accessed 29 November 2016. www.hathitrust.org.

Chicago Distribution Center. Accessed 20 November 2016. http://press .uchicago.edu/cdc.html.

Chritton, Susan. *Personal Branding for Dummies.* 2nd ed. Hoboken, New Jersey: John Wiley & Sons, Inc., 2014.

College Art Association. "Code of Best Practices in Fair Use for the Visual Arts" (2015). Accessed 3 December 2016. http://www.collegeart.org/fair-use/.

Conover, Ted. *Newjack: Guarding Sing Sing.* New York: Knopf Doubleday, 2001.

Copyright Advisory Office, Columbia University Libraries. Accessed 30 November 2016. https://copyright.columbia.edu/about.html.

Copyright & Fair Use, Stanford University Libraries. Accessed 30 November 2016. http://fairuse.stanford.edu/.

Creative Commons. Accessed 3 December 2016. https://search.creativecom mons.org/.

Crews, Kenneth D. *Copyright Law for Librarians and Educators: Creative Strategies and Practical Solutions.* 3rd ed. Chicago: American Library Association, 2012.

Crews, Kenneth D., and Dwayne K. Buttler. "Fair Use Checklist." Accessed 30 November 2016. https://copyright.columbia.edu/basics/fair-use/fair-use -checklist.html.

ebrary. Accessed 20 November 2016. http://proquest.libguides.com/ebrary.

Federal Register: December 6, 2000 (Volume 65, Number 235), 76260–76264. Accessed 13 September 2016. http://ori.hhs.gov/federal-research-misconduct -policy.

Germano, William. *From Dissertation to Book.* 2nd ed. Chicago: University of Chicago Press, 2013.

Getty Museum. Accessed 3 December 2016. http://www.getty.edu/about /opencontent.html.

Hart, Jack. *Storycraft: The Complete Guide to Writing Narrative Nonfiction.* Chicago: University of Chicago Press, 2011.

Herr, Melody. "Communities of American Archaeology." PhD diss., Johns Hopkins University, 2000.

Herr, Melody. "Frontier Stories: Reading and Writing Plains Archaeology." *American Studies* 44 (Fall 2003): 77–98.

Herr, Melody. *Summer of Discovery.* Lincoln: University of Nebraska Press, 2006.

Knowledge Unlatched. Accessed 27 December 2016. http://www.knowl edgeunlatched.org/.

La Raja, Raymond J., and Brian F. Schaffner. *Campaign Finance and Political Polarization: When Purists Prevail.* Ann Arbor: University of Michigan Press, 2015. Accessed 27 December 2016. http://quod.lib.umich.edu/u/ump /13855466.0001.001.

"Lingua Franca" blog, *The Chronicle of Higher Education,* ongoing. Accessed 20 December 2016. http://www.chronicle.com/blogs/linguafranca.

"Little Miss Muffet." Accessed 29 November 2016. http://www.rhymes.org.uk /little_miss_muffet.htm

Machiavelli, Niccoló. *The Essential Writings of Machiavelli.* Translated and edited by Peter Constantine. New York: Random House, Inc., 2007.

Martin, Steve. *Let's Get Small.* Warner Bros. Records, Inc., 1977.

McKeown, Greg. *Essentialism: The Disciplined Pursuit of Less.* New York: The Crown Publishing Group, 2014.

Metropolitan Museum of Art. Open Access for Scholarly Content Policy. Accessed 3 December 2016. http://www.metmuseum.org/about-the-met /policies-and-documents/image-resources.

Miller, Jane E. *The Chicago Guide to Writing about Numbers: The Effective Presentation of Quantitative Information.* 2nd ed. Chicago: University of Chicago Press, 2015.

Mulvany, Nancy C. *Indexing Books.* 2nd ed. Chicago: University of Chicago Press, 2005.

National Book Network. Accessed 20 November 2016. http://nbnbooks.com/.

National Gallery of Art. Accessed 3 December 2016. http://americanhistory.si .edu/collections.

National Museum of American History, Smithsonian Institution. Accessed 3 December 2016. http://americanhistory.si.edu/collections.

National Museum of Natural History, Smithsonian Institution. Accessed 3 December 2016. http://naturalhistory.si.edu/rc/.

"Old McDonald Had a Farm." The British Council, 2012. Accessed 29 November 2016. www.britishcouncil.org/learnenglishkids.

Oxford University Press. "Social Media Author Guidelines." Accessed 24 November 2016. https://global.oup.com/academic/authors/social-media -guidelines.

Poets & Writers. Literary Agents Database. Accessed 24 February 2017. https:// www.pw.org/literary_agents.

Prose, Francine. *Reading Like a Writer: A Guide for People Who Love Books and Those Who Want to Write Them.* New York: HarperCollins Publishers, 2007.

Public Domain Sherpa. Accessed 8 March 2017. http://www.publicdomain sherpa.com/public-domain-maps-resources.html.

Roig, Miguel. *Avoiding Plagiarism, Self-Plagiarism, and Other Questionable Writing Practices: A Guide to Ethical Writing* (2003, revised in 2006 and 2015). Accessed 29 November 2016. http://ori.hhs.gov/avoidingplagiarism-self-plagiarism-and-other-questionablewriting-practices-guide-ethical-writing.

Ruckley, Brian. *Winterbirth, The Godless World Trilogy: Book One*. New York: Orbit, 2006.

Sambuchino, Chuck, ed. *Guide to Literary Agents 2017: The Most Trusted Guide to Getting Published*. New York: F+W Media, 2016. Writers Market, Category Search—Literary Agents. Accessed 24 February 2017. http://www.writers market.com/MarketListings/LiteraryAgents/search.

Sirgo, Henry B. "U.S. Import and Export of Firearms." In *Guns and Contemporary Society: The Past, Present, and Future of Firearms and Firearm Policy*, ed. Glenn H. Utter, Vol. 2, 51–74. Santa Barbara, CA: ABC-CLIO, LLC, 2015.

Sword, Helen. "'Write Every Day!': A Mantra Dismantled." *International Journal for Academic Development* 21:4, 312–322. Published online 11 August 2016. Accessed 30 December 2016. DOI: 10.1080/1360144X.2016.1210153.

Sword, Helen. *The Writer's Diet: A Guide to Fit Prose*. Chicago: University of Chicago Press, 2016. Companion website. Accessed 29 December 2016. http:// writersdiet.com/.

Toor, Rachel. "Held Hostage at a University Press." *The Chronicle of Higher Education*, 9 December 2013. Accessed 28 December 2016. http://www .chronicle.com/article/Held-Hostage-at-a-university/143523?cid=megamenu.

Toor, Rachel. "Scholars Talk Writing." *The Chronicle of Higher Education*, ongoing. Accessed 29 December 2016. http://www.chronicle.com/special report/Scholars-Talk-Writing/26.

Toor, Rachel. "Write a Book and Become an Employee of Your Former Self." *The Chronicle of Higher Education*, 13 October 2014. Accessed 24 November 2016. http://www.chronicle.com/article/Write-a-BookBecome-an/149363.

Turabian, Kate L. *A Manual for Writers of Research Papers, Theses, and Dissertations, Chicago Style for Students & Researchers*. 8th ed. Revised by Wayne C. Booth, Gregory G. Colomb, Joseph M. Williams, and the University of Chicago Press Editorial Staff. Chicago: University of Chicago Press, 2013.

U.S. Copyright Act, 17 *U.S.C.* Accessed 28 December 2016. http://www .copyright.gov/title17/.

U.S. Geological Survey. Accessed 8 March 2017. https://www.usgs.gov/.

U.S. Library of Congress. Accessed 28 December 2016. https://www.loc.gov /collections/.

University of California Press. Accessed 27 December 2016. http://www .ucpress.edu/openaccess.php.

University of California Press. "Author Toolkit." Accessed 23 November 2016. http://www.ucpress.edu/toolkit.php.

University of Chicago Press Staff, eds. *Indexes: A Chapter from The Chicago Manual of Style Sixteenth Edition*. Chicago: University of Chicago Press, 2010.

University of Chicago Press Staff, eds. *The Chicago Manual of Style: The Essential Guide for Writers, Editors, and Publishers*. 16th ed. Chicago: University of Chicago Press, 2010.

Yale University Library Map Collection. Accessed 8 March 2017. http://web .library.yale.edu/maps.

Index

About the Author

MELODY HERR spent more than 16 years in scholarly publishing and acquired more than 250 books for several publishers, including Johns Hopkins University Press and the University of Michigan Press. During the course of her publishing career, she coached authors writing books in a number of fields ranging from regional topics, business history, and U.S. history to law, political science, and international relations. She continues to assist authors in her current position as head of the Office of Scholarly Communications at the University of Arkansas. A writer herself, Melody has published nonfiction and historical fiction for young readers as well as scholarly work.